CLIMATE CHANGE AND INFRASTRUCTURE

DECISION MAKING ISSUES AND ADAPTATION MEASURES

CLIMATE CHANGE AND ITS CAUSES, EFFECTS AND PREDICTION

Additional books in this series can be found on Nova's website under the Series tab.

Additional e-books in this series can be found on Nova's website under the e-book tab.

CLIMATE CHANGE AND ITS CAUSES, EFFECTS AND PREDICTION

CLIMATE CHANGE AND INFRASTRUCTURE

DECISION MAKING ISSUES AND ADAPTATION MEASURES

DUSTIN ETHANS
EDITOR

nova publishers
New York

For permission to use material from this book please contact us:
Telephone 631-231-7269; Fax 631-231-8175
Web Site: http://www.novapublishers.com

Library of Congress Cataloging-in-Publication Data

ISBN: 978-1-63463-061-0

Published by Nova Science Publishers, Inc. † New York

CONTENTS

Preface		**vii**
Chapter 1	Climate Change: Future Federal Adaptation Efforts Could Better Support Local Infrastructure Decision Makers *United States Government Accountability Office*	**1**
Chapter 2	Climate Change Science: Key Points *Jane A. Leggett*	**87**
Index		**107**

PREFACE

According to the National Research Council (NRC) and others, infrastructure such as roads and bridges, wastewater systems, and National Aeronautics and Space Administration (NASA) centers are vulnerable to changes in the climate. Changes in precipitation and sea levels, as well as increased intensity and frequency of extreme events, are projected by NRC and others to impact infrastructure in a variety of ways. When the climate changes, infrastructure—typically designed to operate within past climate conditions—may not operate as well or for as long as planned, leading to economic, environmental, and social impacts. This book examines the impacts of climate change on roads and bridges, wastewater systems, and NASA centers; the extent to which climate change is incorporated into infrastructure planning; factors that enabled some decision makers to implement adaptive measures; and federal efforts to address local adaptation needs, as well as potential opportunities for improvement.

Chapter 1 – The federal government invests billions of dollars annually in infrastructure, such as roads and bridges, facing increasing risks from climate change. Adaptation—defined as adjustments to natural or human systems in response to actual or expected climate change— can help manage these risks by making infrastructure more resilient. GAO was asked to examine issues related to infrastructure decision making and climate change. This report examines (1) the impacts of climate change on roads and bridges, wastewater systems, and NASA centers; (2) the extent to which climate change is incorporated into infrastructure planning; (3) factors that enabled some decision makers to implement adaptive measures; and (4) federal efforts to address local adaptation needs, as well as potential opportunities for improvement. GAO reviewed climate change assessments; analyzed relevant reports; interviewed stakeholders from professional associations and federal agencies; and visited infrastructure projects and interviewed local decision makers at seven sites where adaptive measures have been implemented.

Chapter 2 – Though climate change science often is portrayed as controversial, broad scientific agreement exists on many points: The Earth's climate is warming and changing. Human-related emissions of greenhouse gases (GHG) and other pollutants have contributed to warming observed since the 1970s and, if continued, would tend to drive further warming, sea level rise, and other climate shifts. Volcanoes, the Earth's relationship to the Sun, solar cycles, and land cover change may be more influential on climate shifts than rising GHG concentrations on other time and geographic scales. Human-induced changes are super-imposed on and interact with natural climate variability. The largest uncertainties in climate projections surround feedbacks in the Earth system that augment or dampen the initial influence, or affect the pattern of changes. Feedback mechanisms are apparent in clouds, vegetation, oceans, and potential emissions from soils. There is a wide range of projections of future, human-induced climate change, all pointing toward warming and associated sea level rise, with wider uncertainties regarding the nature of precipitation, storms, and other important aspects of climate. Human societies and ecosystems are sensitive to climate. Some past climate changes benefited civilizations; others contributed to the demise of some societies. Small future changes may bring benefits for some and adverse effects to others. Large climate changes would be increasingly adverse for a widening scope of populations and ecosystems. As is common and constructive in science, scientists debate finer points. For example, a large majority but not all scientists find compelling evidence that rising GHG have contributed the most influence on global warming since the 1970s, with solar radiation a smaller influence on that time scale. Most climate modelers project important impacts of unabated GHG emissions, with low likelihoods of catastrophic impacts over this century. Human influences on climate change would continue for centuries after atmospheric concentrations of GHG are stabilized, as the accumulated gases continue to exert effects and as the Earth's systems seek to equilibrate. The U.S. government and others have invested billions of dollars in research to improve understanding of the Earth's climate system, resulting in major improvements in understanding while major uncertainties remain. However, it is fundamental to the scientific method that science does not provide absolute proofs; all scientific theories are to some degree provisional and may be rejected or modified based on new evidence. Private and public decisions to act or not to act, to reduce the human contribution to climate change or to prepare for future changes, will be made in the context of accumulating evidence (or lack of evidence), accumulating GHG concentrations, ongoing debate about risks, and other considerations (e.g., economics and distributional effects).

In: Climate Change and Infrastructure
Editor: Dustin Ethans
ISBN: 978-1-63463-061-0

Chapter 1

CLIMATE CHANGE: FUTURE FEDERAL ADAPTATION EFFORTS COULD BETTER SUPPORT LOCAL INFRASTRUCTURE DECISION MAKERS*

United States Government Accountability Office

WHY GAO DID THIS STUDY

The federal government invests billions of dollars annually in infrastructure, such as roads and bridges, facing increasing risks from climate change. Adaptation—defined as adjustments to natural or human systems in response to actual or expected climate change— can help manage these risks by making infrastructure more resilient.

GAO was asked to examine issues related to infrastructure decision making and climate change. This report examines (1) the impacts of climate change on roads and bridges, wastewater systems, and NASA centers; (2) the extent to which climate change is incorporated into infrastructure planning; (3) factors that enabled some decision makers to implement adaptive measures; and (4) federal efforts to address local adaptation needs, as well as potential opportunities for improvement.

* This is an edited, reformatted and augmented version of the United States Government Accountability Office publication, GAO-13-242, dated April 2013.

GAO reviewed climate change assessments; analyzed relevant reports; interviewed stakeholders from professional associations and federal agencies; and visited infrastructure projects and interviewed local decision makers at seven sites where adaptive measures have been implemented.

WHAT GAO RECOMMENDS

GAO recommends, among other things, that a federal entity designated by the Executive Office of the President (EOP) work with agencies to identify for local infrastructure decision makers the best available climate- related information for planning, and also to update this information over time. Relevant EOP entities did not provide official comments, but instead provided technical comments, which GAO incorporated, as appropriate.

WHAT GAO FOUND

According to the National Research Council (NRC) and others, infrastructure such as roads and bridges, wastewater systems, and National Aeronautics and Space Administration (NASA) centers are vulnerable to changes in the climate. Changes in precipitation and sea levels, as well as increased intensity and frequency of extreme events, are projected by NRC and others to impact infrastructure in a variety of ways. When the climate changes, infrastructure— typically designed to operate within past climate conditions—may not operate as well or for as long as planned, leading to economic, environmental, and social impacts. For example, the National Oceanic and Atmospheric Administration estimates that, within 15 years, segments of Louisiana State Highway 1— providing the only road access to a port servicing 18 percent of the nation's oil supply—will be inundated by tides an average of 30 times annually due to relative sea level rise. Flooding of this road effectively closes the port.

Decision makers have not systematically considered climate change in infrastructure planning for various reasons, according to representatives of professional associations and agency officials who work with these decision makers. For example, more immediate priorities—such as managing aging infrastructure—consume time and resources, limiting decision makers' ability to consider and implement climate adaptation measures. Difficulties in obtaining and using information needed to understand vulnerabilities and inform adaptation decisions pose additional challenges.

Key factors enabled some local decision makers to integrate climate change into infrastructure planning. As illustrated by GAO's site visits and relevant studies, these factors included (1) having local circumstances such as weather-related crises that spurred action, (2) learning how to use available information, (3) having access to local expertise, and (4) considering climate impacts within existing planning processes. As one example, the Milwaukee Metropolitan Sewerage District managed risks associated with more frequent extreme rainfall events by enhancing its natural systems' ability to absorb runoff by, for instance, preserving wetlands. This effort simultaneously expanded the sewer system's capacity while providing other community and environmental benefits. District leaders enabled these changes by prioritizing adaptation, using available local- level climate projections, and utilizing local experts for assistance.

GAO's report identifies several emerging federal efforts under way to facilitate more informed adaptation decisions, but these efforts could better support the needs of local infrastructure decision makers in the future, according to studies, local decision makers at the sites GAO visited, and other stakeholders. For example, among its key efforts, the federal government plays a critical role in producing the information needed to facilitate more informed local infrastructure adaptation decisions. However, as noted by NRC studies, this information exists in an uncoordinated confederation of networks and institutions, and the end result of it not being easily accessible is that people may make decisions—or choose not to act—without it. Accordingly, a range of studies and local decision makers GAO interviewed cited the need for the federal government to improve local decision makers' access to the best available information to use in infrastructure planning.

ABBREVIATIONS

AASHTO	American Association of State Highway Transportation Officials
CEQ	Council on Environmental Quality
DOD	Department of Defense
EPA	Environmental Protection Agency
ISO	International Organization for Standardization
NASA	National Aeronautics and Space Administration

NEPA	National Environmental Policy Act
NOAA	National Oceanic and Atmospheric Administration
NRC	National Research Council
OMB	Office of Management and Budget
OSTP	Office of Science and Technology Policy
Recovery Act	American Recovery and Reinvestment Act of 2009
RISA	Regional Integrated Science and Assessments
TRB	Transportation Research Board
USGCRP	United States Global Change Research Program

***

April 12, 2013

The Honorable Sheldon Whitehouse
Chairman
Subcommittee on Oversight
Committee on Environment and Public Works
United States Senate

The Honorable Max Baucus
Chairman
Subcommittee on Transportation and Infrastructure
Committee on Environment and Public Works
United States Senate

Extreme weather events and climate change pose risks to physical infrastructure, such as roads and bridges, essential to the economic well- being of the United States. As evident from Superstorm Sandy in October 2012, a single extreme weather event can cause not only extensive loss of life, but also tens of billions of dollars in damages to transportation systems, utilities (e.g., wastewater treatment and collection facilities), buildings, and other critical infrastructure. As noted in a December 7, 2012, memorandum from the Office of Management and Budget (OMB) to the Speaker of the House of Representatives and the Senate Majority Leader requesting funds for Superstorm Sandy recovery efforts,

> To build a more resilient Nation prepared to face both current and future challenges, including a changing climate, federal agencies in

partnership with state, local, and tribal officials, and the science community, should inform all plans for recovery and rebuilding to address the increased risk and vulnerabilities of extreme weather, sea level rise, and coastal flooding.

Typically, climate change is described in terms of average annual changes in temperature or precipitation, but it is also associated with shifts in the frequency and severity of extreme weather events to which physical infrastructure is particularly vulnerable. These vulnerabilities are illustrated by the 11 extreme weather events—including Superstorm Sandy—that each caused more than $1 billion in damages in 2012, according to the National Oceanic and Atmospheric Administration's (NOAA) National Climatic Data Center.[1] In 2011, the United States experienced 14 extreme weather and climate events that cost more than $1 billion in losses, according to NOAA, including Hurricane Irene, which also caused flooding across the Northeast.

As observed by the United States Global Change Research Program (USGCRP), the impacts and costliness of these disasters—resulting from floods, drought, and other events such as tropical cyclones—will increase in significance as what are considered "rare" events today become more common and intense due to climate change.[2] In addition, USGCRP's 2009 National Climate Assessment found that climate-related changes— such as rising temperature and sea level—will exacerbate the effects of other challenges we face as a nation, such as air pollution; population growth; urbanization; and other social, economic, and environmental stresses.[3] According to the National Research Council (NRC), USGCRP, and others, greenhouse gases already in the atmosphere will continue altering the climate system into the future, regardless of emissions control efforts.[4] Therefore, climate change adaptation—defined here as adjustments to natural or human systems in response to actual or expected climate change—is an important part of the response to climate change.[5]

As we reported in October 2009, policymakers are increasingly viewing climate change adaptation as a risk management strategy to protect vulnerable sectors and communities that might be affected by changes in the climate.[6] We also reported that federal, state, and local agencies were beginning to take action. Since our 2009 report, many agencies within the federal government have developed adaptation initiatives, including activities focused on infrastructure such as roads and bridges, wastewater management systems, and large federal facilities like National Aeronautics and Space Administration (NASA) centers. Further, in its 2010 Quadrennial Defense Review, the

Department of Defense (DOD) recognized the national security risk posed by climate change, noting that while climate change alone does not cause conflict, it may act as an accelerant of instability or conflict, placing a burden to respond on civilian institutions and militaries around the world.[7] The review also acknowledged that extreme weather events may lead to increased demands for defense support to civil authorities for humanitarian assistance or disaster response. The Department of Homeland Security's 2010 Quadrennial Homeland Security Review similarly recognized that America's national interests are threatened by global challenges and long-term trends including climate change.[8] Because of these and other concerns, in February 2013, we added *Limiting the Federal Government's Fiscal Exposure by Better Managing Climate Change Risks* to our High Risk List.[9]

In this context, you asked us to review the consequences of climate change on U.S. infrastructure. Specifically, this report examines (1) what is known about the impacts of climate change on the nation's infrastructure, specifically roads and bridges, wastewater management systems, and NASA centers; (2) the extent to which potential climate change impacts are incorporated into infrastructure planning; (3) examples in which climate change impacts were integrated into infrastructure planning; and the factors that enabled some decision makers, such as road engineers, to do so; and (4) federal efforts to address the adaptation needs of local infrastructure decision makers and potential opportunities for improvement.

To examine what is known about the impacts of climate change on the nation's infrastructure, we reviewed assessments from the NRC and USGCRP.[10] We selected the road and bridge and wastewater management system infrastructure categories because they are supported by significant federal funding and are the focus of specific federal adaptation initiatives.[11] We selected NASA centers because these facilities are large and manage mission critical assets that are difficult, if not impossible, to move or replace; also, NASA has an emerging effort focused on considering climate change information within the planning for its centers. NASA centers are also instructive examples because they incorporate roads, bridges, wastewater systems, and other infrastructure in one place as a system to support a mission. NASA scheduled climate change adaptation workshops at two of its centers (Langley Research Center and Johnson Space Center) during the time frame of our review. We attended the workshops and collected information from a variety of federal and local stakeholders, including government officials and academic institutions.

To examine the extent to which climate change impacts are incorporated into infrastructure planning we (1) reviewed relevant laws, regulations, and planning guidance; (2) analyzed relevant studies and government reports; and (3) interviewed knowledgeable stakeholders including representatives from, for example, professional associations such as the American Association of State Highway and Transportation Officials. To identify relevant studies and stakeholders, we reviewed our prior climate change work and conducted a literature search and review.

To examine how climate change has been considered in U.S. infrastructure planning, we visited a nonprobability sample of seven selected locations where decision makers had undertaken such planning—three locations focused on roads and bridges (Washington State Route 522; Interstate-10 Twin Span Bridge near New Orleans, Louisiana; and Louisiana State Highway 1) , two locations focused on wastewater management systems (King County Wastewater Treatment Division in Washington and the Milwaukee Metropolitan Sewerage District in Wisconsin), and two NASA centers (Johnson Space Center in Houston, Texas, and Langley Research Center in Hampton, Virginia).[12] To select the specific locations, we reviewed studies, interviewed knowledgeable agency officials, and analyzed Internet-based adaptation case study databases maintained by academic institutions to identify examples where climate change was considered in infrastructure planning, taking into account geographic and climate-impact diversity.

To analyze federal efforts to meet the adaptation needs of local infrastructure decision makers and to identify opportunities for improvement, we (1) interviewed federal officials from the Council on Environmental Quality (CEQ), Department of Transportation's Federal Highway Administration, Environmental Protection Agency (EPA), and USGCRP and (2) reviewed available studies on federal adaptation efforts. Appendix I presents a more detailed description of our scope and methodology.

We conducted this performance audit from October 2011 to April 2013 in accordance with generally accepted government auditing standards. Those standards require that we plan and perform the audit to obtain sufficient, appropriate evidence to provide a reasonable basis for our findings and conclusions based on our audit objectives. We believe that the evidence obtained provides a reasonable basis for our findings and conclusions based on our audit objectives.

BACKGROUND

Policymakers are increasingly viewing adaptation as a risk management strategy to protect vulnerable infrastructure that might be affected by changes in the climate. While adaptation measures—such as raising river or coastal dikes to protect infrastructure from sea level rise, building higher bridges, or increasing the capacity of stormwater systems—may be costly, there is a growing recognition that the cost of inaction could be greater. As stated in a 2010 NRC report, even though there are still uncertainties regarding the exact nature and magnitude of climate change impacts, mobilizing now to increase the nation's adaptive capacity can be viewed as an insurance policy against climate change risks.[13] In this context, it is important to understand (1) federal infrastructure investment, (2) the condition of existing infrastructure, (3) climate change adaptation as a risk management tool, and (4) the limited federal role in planning infrastructure projects.

Federal Infrastructure Investment

In total, the United States has about 4 million miles of roads; 30,000 wastewater treatment and collection facilities; and over 800,000 federal facilities such as military installations that provide for the nation's defense and facilities where complex scientific and technological research is conducted. Collectively, this infrastructure connects communities, protects public health and the environment, and facilitates trade and economic growth, among other important functions. The nation's highway and wastewater infrastructure is primarily owned and operated by state and local governments and the private sector. For instance, state and local governments own about 98 percent of the nation's bridges.[14]

The federal government spends billions of dollars every year on transportation and wastewater infrastructure through a variety of funding mechanisms. According to a 2010 Congressional Budget Office report, total public spending on transportation and water infrastructure exceeds $300 billion annually, with roughly 25 percent of this amount coming from the federal government and the rest coming from state and local governments.[15] For the most part, the federal government supports these infrastructure investments through federal assistance to states and local communities. For example, EPA's Clean Water State Revolving Fund—a federal program that provides states and local communities with independent and sustainable sources of financial assistance, such as low- or no-interest loans to fund water quality projects—received an appropriation of just over $1.4 billion in fiscal year 2012.[16] From 1956 to 2007, the largest portion of annual public funding for transportation and water infrastructure was dedicated to highways.

The federal government also owns and manages certain types of infrastructure. According to the General Services Administration, the federal government's portfolio of building assets totaled approximately 3.35 billion square feet of space and over 800,000 building and structural assets with a total operating cost of $30.8 billion in 2010.[17] This total includes large federal complexes such as NASA centers where scientists and engineers conduct research, design new aerospace technologies, and operate the International Space Station, among other activities.

NASA's real property holdings include more than 5,000 buildings and other structures such as wind tunnels, laboratories, launch pads, and test stands, according to a December 2011 NASA Inspector General report.[18] In total, these NASA assets represent more than $32 billion in current replacement value.

Condition of Existing Infrastructure

The infrastructure examined for this report—roads, bridges, wastewater management systems, and NASA centers—was designed to last for decades. More specifically, according to the American Society of Civil Engineers, the average bridge in the United States is designed to last 50 years, and EPA data indicate that wastewater treatment plants typically have an expected useful life of 20 to 50 years before they require expansion or rehabilitation.[19] Over 80 percent of NASA's facilities are more than 40 years old and reaching the end of their designated life spans.[20]

Our past work and other studies have reported that much of the nation's physical infrastructure is in poor condition and in need of repair or replacement. For example, as we reported in June 2010, many wastewater management systems were constructed more than 50 years ago and are reaching the end of their useful lives.[21] Many of these systems do not have the capacity to treat increasing volumes of wastewater, particularly during periods of wet weather, leading to the release of untreated wastewater into water bodies. Citing such concerns, the American Society of Civil Engineers 2009 Report Card for America's Infrastructure graded the condition of the nation's wastewater infrastructure as "D-," or between "poor" and "failing" on their rating scale.[22] Roads and bridges fared similarly in the report card, also earning a "D-," and a "C" for "mediocre," respectively, due to identified structural deficiencies and other factors.

Estimates to repair, replace, or upgrade aging roads, bridges, wastewater management systems, and federal facilities are in the hundreds of billions of dollars, not accounting for any additional costs to be incurred due to a changing climate, an important consideration for proposed adaptive measures that may require significant redesign, retrofitting, or replacement of planned or existing infrastructure in response to a changing climate. As we reported in May 2008, the current fiscal environment makes it even more important that federal, state, and local governments make prudent decisions on how to invest limited available resources as they address infrastructure needs.[23] Yet, in many cases, we reported that federal infrastructure investment decisions are still based on conditions, priorities, and approaches that were established decades ago and are not well suited to addressing complex, crosscutting, and emerging challenges like climate change. Our May 2008 report identified principles that could guide a reexamination of federal infrastructure programs. These principles include creating well-defined goals based on identified areas of national interest, establishing and clearly defining the federal role in achieving each goal, incorporating performance and accountability into funding decisions, and employing the best tools and approaches to emphasize return on investment.

Climate Change Adaptation as a Risk Management Tool

Climate change adaptation addresses the vulnerability of natural and human systems to changes in the climate and focuses on reducing the damage resulting from those changes.[24] One way to reduce the potential impacts of climate change is to invest in enhancing resilience. As defined by the National Academies, resilience is the ability to prepare and plan for, absorb, recover from, and more successfully adapt to adverse events.[25] Enhanced resilience results from better planning to reduce losses, rather than waiting for an event to occur and paying for recovery afterward.

Adaptation efforts are meant to reduce the vulnerability of systems that have some risk of an extreme event or long-term change in conditions. As summarized in the 2010 NRC study,

> America's climate change adaptation choices involve deciding how to cope with climate changes that we cannot, or do not, avoid so that possible disruptions and damages to society, economies, and the environment are minimized and—where possible—so that impacts are converted into opportunities for the country and its citizens. In some cases, such as in Alaska, the need to adapt has already become a reality. In most cases, however, adapting today is about reducing vulnerabilities to emerging or future impacts that could become seriously disruptive if we do not begin to identify response options now; in other words, adaptation today is essentially a risk management strategy.[26]

Further, as we reported in 2009, given the complexity and potential magnitude of climate change and the lead time needed to adapt, preparing for these impacts now may reduce the need for far more costly steps in the decades to come.[27] Of particular importance are planning decisions involving physical infrastructure, which require large capital investments and which, by virtue of their expected life span, will have to be resilient to changes in climate for many decades. Substitutes for infrastructure could also affect adaptation decisions; damages from disruptions due to climate change would be greater, all else equal, when fewer alternatives are available. The long lead time and long life of large infrastructure investments require planning decisions to be made well before further climate change effects are discernible.

Risk management is not a new concept, and it is used extensively almost anywhere decision makers are faced with incomplete information or unpredictable outcomes that may have negative impacts.[28] Broadly defined, risk management is a strategic process for helping policymakers make decisions about assessing risk, allocating finite resources, and taking actions under conditions of uncertainty. Leading risk management guidance recommends a sequence of activities similar to the one described in the International Organization for Standardization (ISO) 31000:2009 standards on risk management.[29] Specifically, these standards recommend that organizations such as federal agencies develop, implement, and continuously improve a framework for integrating risk management into their overall planning, management, reporting processes, and policies.[30] For risk management to be effective, these standards state that an organization should at all levels comply with the following principles:

- *Risk management is not a stand-alone activity that is separate from the main activities and processes of the organization.* Risk management is part of the responsibilities of management and an integral part of all organizational processes, including strategic planning and all project and change management processes.
- *Risk management is part of decision making.* Risk management helps decision makers make informed choices, prioritize actions, and distinguish among alternatives.
- *Risk management explicitly addresses uncertainty.* Risk management explicitly takes account of uncertainty, the nature of that uncertainty, and how it can be addressed.
- *Risk management is based on the best available information.* The inputs to the process of managing risk are based on information sources such as historical data, experience, stakeholder feedback, observation, forecasts, and expert judgment. However, decision makers should inform themselves of, and should take into account, any limitations of the data or modeling used or the possibility of divergence among experts.

Concerning risk management for federal facilities, OMB issues guidance for agencies on managing non-information technology capital assets that contains risk management criteria.[31] Under OMB's guidance, agencies are to complete a business case for physical infrastructure investment that includes sections on alternatives analysis and risk management. Risk must be actively managed throughout the life cycle of the investment, and a risk management plan must be available to OMB upon request.

Limited Federal Role in Planning Infrastructure Projects

The federal government has an inherently limited role in the project-level planning processes central to adapting infrastructure to climate change because these are typically the responsibility of state and local governments—except when federal assets are involved. State and local authorities are primarily responsible for prioritizing and supervising the implementation of water and highway infrastructure projects; therefore, the federal role in these processes is limited.[32]

As specified by law, federal programs for funding roads, bridges, and wastewater infrastructure generally operate as formula grants or similar mechanisms with few explicit requirements to consider climate change in infrastructure projects. For example, federal funding for highways is provided to the states mostly through a series of formula grant programs collectively known as the federal-aid highway program.[33] As we have reported, the Federal Highway Administration has faced challenges in ensuring that federal funds are efficiently and effectively used because the highway program is one in which there is limited federal control—it is a state-administered, federally assisted program.[34] Funds are largely apportioned by formula, and the states enjoy broad flexibility in deciding which projects are supported. Furthermore, for nearly half of federal-aid highway funds, the Federal Highway Administration's responsibility to oversee the design and construction of projects has been assumed by the states. Similarly, EPA officials told us that their ability to influence states to adapt to climate change through the Clean Water State Revolving Fund is limited and that each state is responsible for administering its own revolving funds.

However, certain federal infrastructure programs may begin to consider adaptation in their project-level planning activities. For example, the Moving Ahead for Progress in the 21st Century Act—which was signed into law on July 6, 2012, and authorized over $105 billion in appropriations for surface transportation programs for fiscal years 2013 and 2014—authorizes federal funding to be used for bridge and tunnel projects that protect against extreme events.[35]

Another example is funding appropriated in the American Recovery and Reinvestment Act of 2009 (Recovery Act) for EPA's Green Project Reserve under the Clean Water State Revolving Fund and EPA's Drinking Water State Revolving Fund programs.[36] The Recovery Act appropriated $4 billion for the Clean Water State Revolving Fund and required that not less than 20 percent of these funds—if there are sufficient eligible project applications—be used for projects to address green infrastructure or other environmentally innovative activities.[37] According to EPA, this requirement, known as the Green Project Reserve, has funded projects that facilitate adaptation of clean water facilities to climate change, including green infrastructure and other climate-related and environmentally innovative activities, such as water and energy conservation.[38]

Table 1. Current and Projected Climate Changes in the United States and Impacts to Infrastructure

Category	Current and projected climate changes	Examples of impacts to infrastructure
Temperature	• U.S. average annual temperature has risen more than 2 degrees Fahrenheit over the past 50 years and is projected to rise more in the future—how much depends on the amount of heat-trapping gases emitted globally and how sensitive the climate is to those emissions.	Damage to materials such as more potholes in roads.
Precipitation	• Precipitation has increased an average of about 5 percent over the past 50 years. Projections of future precipitation generally indicate that northern areas will become wetter and southern areas, particularly in the West, will become drier. • The amount of rain falling in the heaviest downpours has increased about 20 percent in the past century, and this trend is very likely to continue, with the largest increases in the wettest places. • In most regions of the country, the fraction of precipitation falling as rain as opposed to snow has increased in the last 50 years.	Flooding and direct damage resulting in sewer overflows.
Extreme weather events and storms	• Many types of extreme weather events, such as heat waves and regional droughts, have become more frequent and intense during the past 40 to 50 years. • The destructive energy of Atlantic hurricanes has increased in recent decades. The intensity of these storms is likely to increase in this century and along with it, associated wind, precipitation, and storm surges. • In the eastern Pacific, the strongest hurricanes have become stronger since the 1980s, even while the total number of storms has decreased. • Cold season storm tracks are shifting northward, and the strongest storms are likely to become stronger and more frequent.	Flooding and direct damage such as the failure of bridges.
Sea level	• Sea level has risen along most of the U.S. coast over the last 50 years and will likely rise more in the future.[a]	Coastal flooding and inundation.

Sources: USGCRP's 2009 National Climate Assessment and NRC's America's Climate Choices: Adapting to the Impacts of Climate Change, 2010.

[a] In December 2012, NOAA released global sea level rise scenarios that helped inform the third National Climate Assessment.

Roads and Bridges, Wastewater Management Systems, and NASA Centers Are Vulnerable to Changes in the Climate

According to NRC and USGCRP assessments, changes in the climate have been observed in the United States and its coastal waters and are projected to grow in severity in the future, thereby increasing the vulnerability of infrastructure such as roads and bridges, wastewater management systems, and NASA centers. As shown in table 1, changes in the climate—including warmer temperatures, changes in precipitation patterns, more frequent and intense storms and extreme weather events, and sea level rise—affect roads and bridges, wastewater management systems, and NASA centers in a variety of ways, according to NRC and USGCRP.

Infrastructure is typically designed to withstand and operate within historical climate patterns. However, according to NRC, as the climate changes and historical patterns—in particular, those related to extreme weather events—no longer provide reliable predictions of the future, infrastructure designs may underestimate the climate-related impacts to infrastructure over its design life, which can range as long as 50 to 100 years.[39] These impacts can increase the operating and maintenance costs of infrastructure or decrease its life span, or both, leading to social, economic, and environmental impacts.

The vulnerability of infrastructure to changes in the climate varies by category and location, as illustrated by our seven site visits, examples from additional interviews we conducted, and assessments we reviewed focused on three infrastructure categories—roads and bridges, wastewater infrastructure, and NASA centers.

Climate Change Impacts on Roads and Bridges

Climate change will have a significant impact on the nation's roads and bridges, according to assessments by NRC, USGCRP, and others. Transportation infrastructure is vulnerable to extremes in precipitation, temperature, and storm surges, which can damage roads, bridges, and roadway drainage systems. For example, if soil moisture levels become too high with increased precipitation, the structural integrity of already aging roads, bridges, and tunnels could be compromised.[40] In addition, USGCRP's 2009 assessment notes that increased precipitation is likely to increase weather-related accidents, delays, and traffic disruptions in a transportation network already challenged by increasing congestion. Evacuation routes are likely to experience increased flooding, and more precipitation falling as rain rather than snow in winter and spring is likely to increase the risk of landslides, slope failures, and floods from the runoff, causing road closures, as well as the need for road and bridge repair and reconstruction. According to technical comments from EPA, increased precipitation could also overwhelm roadside stormwater systems, causing flooding of homes and businesses.

Increases in temperature extremes are projected to generate more freeze-thaw conditions, creating potholes on road and bridge surfaces and resulting in load restrictions on certain roads to minimize damage, according to a 2008 NRC study.[41] In addition, longer periods of extreme heat may compromise pavement integrity by softening asphalt and increasing rutting (i.e., sunken tracks or grooves made by the passage of vehicles).

Storm surge, combined with sea level rise, is projected to generate a wide range of negative impacts on roads and bridges. For example, according to the 2008 NRC study, storm surges are projected to increasingly inundate coastal roads, cause more frequent or severe flooding of low- lying infrastructure, erode road bases, and "scour" bridges by eroding riverbeds and exposing bridge foundations.[42] From an operational perspective, increased storm surges are projected to cause more frequent travel interruptions, especially in low-lying and coastal areas, and necessitate more frequent evacuations, according to the study. The following are specific examples of the observed and projected effects of climate change on roads and bridges from the sites we visited.

Washington State Route 522

Washington state transportation officials told us that they expect that Washington State Route 522, about 35 miles northeast of Seattle, Washington, will be vulnerable to hydrologic changes resulting from changing temperatures and precipitation. More specifically, as the climate warms and glaciers melt, they expect increased sediment loads in nearby waterways. In addition, changes in rain and snow patterns are expected to alter river flows, which have already caused problems at Washington State Route 522's Snohomish River Bridge, located in a vulnerable position downstream of the convergence of the flash flood prone Skykomish River and the slower-moving Snoqualmie River. Due to flash flooding in this river system, state transportation officials said that they have had to repair scour damage at the Snohomish River Bridge. When designing the new project to replace the bridge and widen this 4-mile stretch of Washington State Route 522, transportation officials told us that they anticipated that hydrologic changes would continue to pose scour risks to the bridge.

Interstate 10 Twin Span Bridge and Louisiana State Highway 1

The two other road and bridge locations we visited highlight their vulnerability to storms and relative sea level rise—the combination of global sea level rise and changes in land surface elevation resulting from land loss through subsidence, or the sinking of land that can lead to submergence. Specifically, the Interstate 10 Twin Span Bridge, which crosses Lake Pontchartrain near New Orleans, and the southern portion of Louisiana State Highway 1 are both located in the low-lying central Gulf Coast region. This region is already prone to flooding during heavy rainfall events, hurricanes, and tropical storms, and USGCRP assessments expect that the region will become increasingly susceptible to inundation as barrier islands erode and subside into the Gulf of Mexico. In its 2008 study, USGCRP estimated that the region could experience as much as 6 to 7 feet of relative sea level rise in Louisiana and East Texas, an area home to a dense network of transportation assets.[43] According to this study, the "middle range" of potential sea level rise (2 to 4 feet) indicates that a vast portion of the Gulf Coast from Houston to Mobile may be inundated over the next 50 to 100 years.

The Twin Span Bridge has already been damaged by one extreme weather event—Hurricane Katrina. In 2005, Hurricane Katrina generated a large storm surge across Lake Pontchartrain, lifting many of Twin Span's 255-ton concrete bridge spans off of their piers, as shown in figure 1.

Source: Volkert Construction Services, Inc.

Figure 1. Damaged Segments of the Twin Span Bridge.

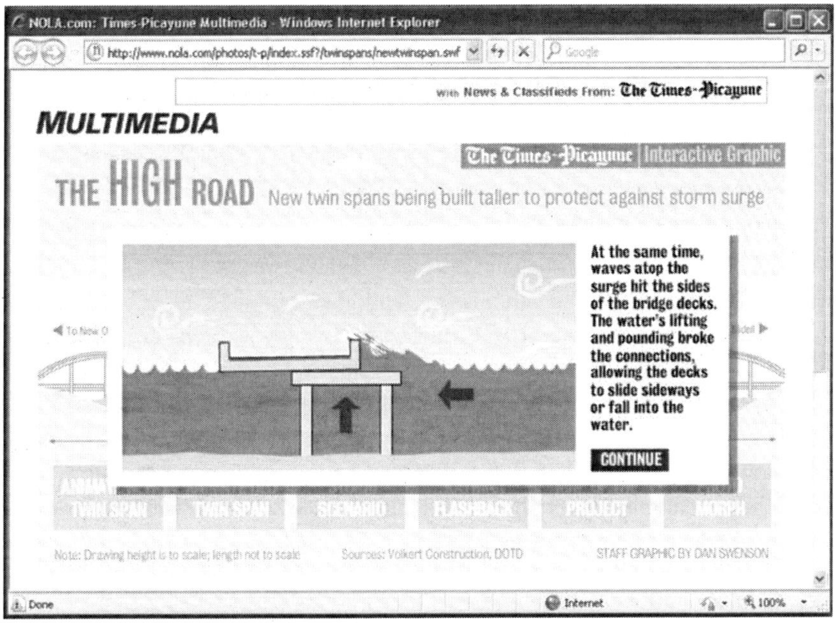

Figure 2. Interactive Graphic Illustrating the Impact of Hurricane Katrina Storm Surge and Waves on I-10 Twin Span Bridge.

Some of the spans toppled into the lake while others were seriously misaligned. The interactive graphic figure 2, below, illustrates how storm surge combined with wind-driven waves to knock the spans off their piers. Visit http://www.nola.com/photos/t-p/index.ssf?/twinspans/newtwinspan.swf to activate the graphic in a Web browser on your computer, and then select the "Katrina Flashback" box to access the animation.

The sections of Louisiana State Highway 1 we visited are also in a particularly vulnerable location near the Gulf of Mexico, according to locally based federal and state officials. The highway provides the only road access to Port Fourchon, which services virtually all deep-sea oil operations in the Gulf of Mexico, and the Louisiana Offshore Oil Port, the nation's only deepwater oil port capable of unloading very large crude carriers. Collectively, Louisiana State Highway 1 currently supports 18 percent of the nation's oil supply. Flooding of this road effectively closes the port. According to NOAA officials, relative sea level rose an average of about 0.4 inches annually from 1947 to 2006 at a tidal gauge in nearby Grand Isle, LA. This is equivalent to a change of approximately 3 feet in 100 years, which a NOAA official described as one of the highest rates of relative sea level rise in the world. Currently, Louisiana State Highway 1 is closed an average of 3.5 days annually due to inundation. However, within 15 years, NOAA anticipates that the at-grade portions of Louisiana State Highway 1 will be inundated by tides an average of 30 times annually even in the absence of extreme weather.

Because of Port Fourchon's significance to the national, state, and local oil industry, the U.S. Department of Homeland Security, in July 2011, estimated that a closure of 90 days could reduce national gross domestic product by $7.8 billion.[44] In addition to these anticipated economic impacts, local officials also said that they are concerned about the safety of area residents and workers who rely on Louisiana State Highway 1 as their sole evacuation route during extreme weather events. Workers travelling between the port and their homes must navigate a low-lying segment of Louisiana State Highway 1, parts of which were built 4 feet above sea level in an area where current high tide levels are 2.5 feet above sea level. Figure 3 shows Louisiana State Highway 1 leading to Port Fourchon.

Source: NOAA.

Figure 3. Louisiana State Highway 1 Leading to Port Fourchon.

Climate Change Impacts on Wastewater Management Systems

Climate change will have a significant impact on the nation's wastewater management infrastructure—including treatment plants and wastewater collection systems, according to studies from wastewater professional associations and EPA, and an assessment from USGCRP.[45] Representatives from the National Association of Clean Water Agencies and EPA officials we interviewed said that the most direct impacts of climate change involve more frequent flooding and damage to wastewater infrastructure. Climate changes that alter the local hydrology—such as sea level rise, especially when combined with higher storm surges, increased precipitation amounts, or more frequent and intense downpours—can cause increased flooding and inundation of wastewater management infrastructure, according to the professional association study and USGCRP's 2009 National Climate Assessment. Stronger storms, which USGCRP projects in some locations, may exacerbate these impacts. Wastewater infrastructure is particularly vulnerable to climate change impacts because it is commonly built in low-lying areas near a body of water and because it is designed for historically observed hydrologic conditions that may not be as relevant for future scenarios.

Some locations could experience other, less direct, climate change impacts from higher temperatures or drought conditions that alter the characteristics of the wastewater flowing into a treatment plant—for example, by concentrating pollutants or increasing water temperatures— thereby reducing the effectiveness of a system's treatment processes that were designed for different characteristics. In addition, treatment plants may need to adopt alternative strategies to managing discharge of treated or partially treated effluent if the condition of receiving water is altered by climate impacts, according to technical comments from EPA. For example, according to EPA's comments, the flow of the receiving water body may be too low to dilute discharge enough to meet water quality standards. Climate change impacts to wastewater management systems can increase treatment costs, increase maintenance and replacement costs, and compromise biological treatment systems resulting in impaired water quality. In the worst cases, according to EPA officials, climate change impacts could cause a system to fail, creating risks to public health.

Potential climate change impacts on wastewater management systems are not limited to coastal areas, since changes in precipitation and extreme events could affect wastewater management systems across the country. According to USGCRP's 2009 National Climate Assessment, the amount of rain falling in the heaviest downpours has increased approximately 20 percent on average in the past century, and this trend is very likely to continue, with the largest increases in the wettest places. During the past 50 years, the greatest increases in heavy precipitation occurred in the Northeast and the Midwest. Besides flooding and related storm damage at treatment plants, increased precipitation creates problems for combined and separate sewer systems that collect and carry sewage to treatment facilities. Specifically, these precipitation changes can increase the amount of runoff, which by design combines with sewage in a combined sewer system, and can lead to increased infiltration and inflow into aging separated systems. These increases can overwhelm the capacity of sewer systems, causing overflows that bypass treatment and result in the discharge of untreated wastewater into receiving water bodies.

Wastewater management systems are typically designed to provide a specific level of service based on a number of design factors that include a particular storm frequency, duration, and intensity. For example, according to one set of commonly used design standards, treatment plant components are typically designed for 25-, 50-, or 100-year storms.[46] Changes in characteristics of strong storms—for instance, a storm that historically occurred once every 100 years may occur every 50 years in the future—could cause wastewater management systems to be overwhelmed more frequently. Climate change impacts have added to existing stresses—including aging infrastructure and urbanization—that already tax the capacities of many of the country's wastewater management systems and challenge communities' ability to pay for them. Specific impacts that have been observed in the two locations we visited are discussed in the following sections.

King County, Washington, Wastewater Treatment Division

In some cases, flooding and saltwater intrusion already affect wastewater management systems. For example, the wastewater infrastructure of King County, Washington (which includes metropolitan Seattle) has been affected by saltwater intrusion into its combined sewer system.[47] Such incidents can compromise the biological treatment system of the wastewater treatment plant. Officials from the King County Wastewater Treatment Division noted that higher sea level will exacerbate the issue and affect the ability of the system to operate as designed and cause excessive corrosion to the facilities. To evaluate this possibility, King County performed a vulnerability study in 2008 to identify the facilities at risk of potential flooding due to sea level rise and storm surges. The study used different regional sea level rise scenarios developed by scientists at the University of Washington, combined with historical tide heights and storm events to develop different scenarios of future tide heights. These tide height scenarios were combined with the elevations of King County's system facilities to identify those at risk of onsite flooding. As shown in figure 4, King County has many facilities—including treatment plants, regulator stations, pump stations, and other components—in tidally influenced areas. The lowest of these facilities—Barton Pump Station, 8th Avenue Regulator Station, Brightwater Flow Meter Vault and Sampling Facility, and Elliott West Combined Sewer Overflow Treatment Plant—lie less than 15 feet above sea level.

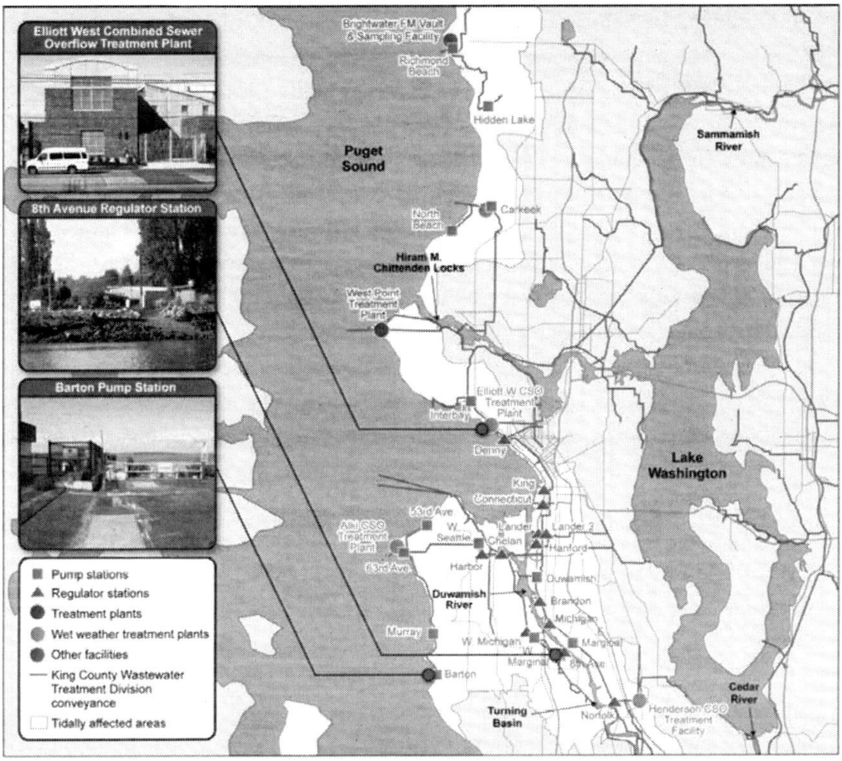

Source: King County Wastewater Treatment Division (map); King County (photo); and GAO.

Figure 4. King County Wastewater Treatment Division Facilities Potentially Vulnerable to Sea Level Rise.

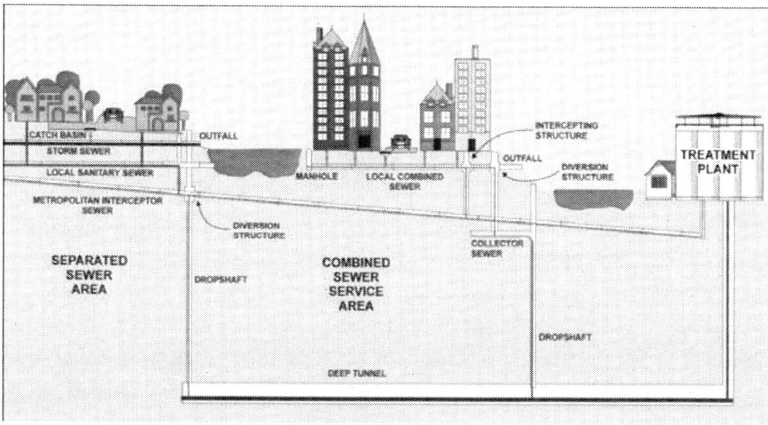

Source: Milwaukee Metropolitan Sewerage District.

Figure 5. Milwaukee Metropolitan Sewerage District's Sewer System Infrastructure.

The 2008 vulnerability study concluded that more than 30 major facilities in King County are at varying levels of risk from sea level rise and storm surge, depending on the rate at which the rise occurs and the probability of an extreme storm event. For example, according to the study, the Barton Pump Station, 8th Avenue Regulator Station, and Brightwater Flow Meter Vault and Sampling Facility—all of which have an elevation of 13 feet—are projected by be flooded every 2 years by 2050 under a high sea level rise scenario (approximately 1.8 feet).

Milwaukee Metropolitan Sewerage District

Due to past problems with sewer overflows, the Milwaukee Metropolitan Sewerage District in Wisconsin significantly increased the capacity of its sewer system.[48] As shown in figure 5, the district completed a $3 billion dollar project in 1993 that included construction of a "deep tunnel" to add additional wastewater storage capacity to its combined and separated sewer systems and decrease the likelihood of combined sewer overflows. In the past, according to Milwaukee Metropolitan Sewerage District officials, this project and the district's other sewer system design decisions were based on a 64-year historical rainfall record from 1940 to 2004. These officials stated that the Milwaukee region's robust sewer infrastructure helps make its system less vulnerable to changes in precipitation that may result from climate change.

However, during our site visit, Milwaukee Metropolitan Sewerage District officials stated that even this more robust infrastructure is vulnerable to projected changes in the climate. In recent years, the Milwaukee region has experienced several extreme precipitation events and, in 2011, scientists at the University of Wisconsin projected that these types of storms will become more common in the future.[49] Specifically, the scientists projected that storm frequency and intensity will increase in early spring, a time during which the sewers are more vulnerable to overflows due to frozen ground conditions that limit infiltration and cause more runoff. Increases in spring precipitation associated with climate change could exceed the capacity of the system and increase the volume and frequency of sewer overflows in the Milwaukee region by mid- century, according to the scientists.

Table 2. Examples of How NASA Centers Are Vulnerable to Observed and Projected Climate Change Impacts

Observed or projected climate changes	Potential climate change impacts on NASA centers
Increased temperatures	• Increased cooling costs in the summer, decreased heating costs in the winter • Changes in plant and animal cycles, including pest and disease vector species • Potential for damage to infrastructure materials • Potential for limiting work and outdoor recreation • Increased health problems related to heat stress
Sea-level rise	• Exacerbated flooding from storm surges • Reduced emergency response capabilities • Increased salinity impacts to drinking water resources and habitats
Increase in precipitation amount and intensity	• Increased flooding from extreme precipitation events • Overloading of stormwater management system • Habitats affected by fluctuating groundwater levels
Increased hurricane intensity/storm surge	• Damage to infrastructure • Changes in shoreline habitats • Reduced emergency response capabilities

Sources: GAO analysis based on NASA studies and NRC and USGCRP assessments.

Climate Change Impacts on NASA Centers

As presented in table 2, NASA centers are vulnerable to climate change in several respects, but potential impacts vary depending upon geographic location.

NASA's centers and associated sites each have different missions and geographic characteristics that affect their vulnerability to climate change. As shown in figure 6, many of NASA's field centers and component sites are near an ocean shoreline. In fact, over two-thirds of NASA's constructed real property value (about $20 billion) is within 16 feet of sea level, according to a 2012 NASA climate change presentation.

NASA is developing the institutional capacity to identify the risks posed to its centers by climate change through a series of multiday climate risk workshops, including two we attended in September 2011 and March 2012. The workshops are intended to, among other functions, share climate information specific to each center with agency officials— including headquarters officials, center leadership, and center managers responsible for overarching "systems" that support mission and operations, such as the center's electrical distribution network—and community stakeholders such as local planning officials. Through these workshops, NASA climate scientists and center personnel have assembled site specific observed and projected changes in the climate for selected centers, and have begun grappling with potential climate impacts on these facilities. We describe two centers we visited—Johnson Space Center and Langley Research Center—in more detail in the following sections, as well as selected emerging efforts within DOD, which has several facilities in close proximity to Langley Research Center:

Johnson Space Center

According to NASA documents obtained at the March 2012 workshop, Johnson Space Center leads NASA's flight-related scientific and medical research efforts, and its professionals direct the development, testing, production, and delivery of U.S. human spacecraft and human spacecraft-related functions, including training space explorers from the United States and Space Station partner nations, including International Space Station crews.[50] As shown in figure 7, the center is located on nearly 1,700 acres in Houston, Texas, near Galveston Bay and the Gulf of Mexico. Ellington Field, part of Johnson Space Center, lies northwest of the center.

Source: NASA and Map Resources (map).

Note: According to NASA's Facility Management Plan, total NASA sites include approximately 63 in the continental United States and 26 overseas, including sites where NASA owns facilities but not the land. NASA's workforce, land holdings, and constructed value are virtually all at the sites shown in this diagram, according to NASA officials.

Figure 6. NASA Field Centers and Component Sites.

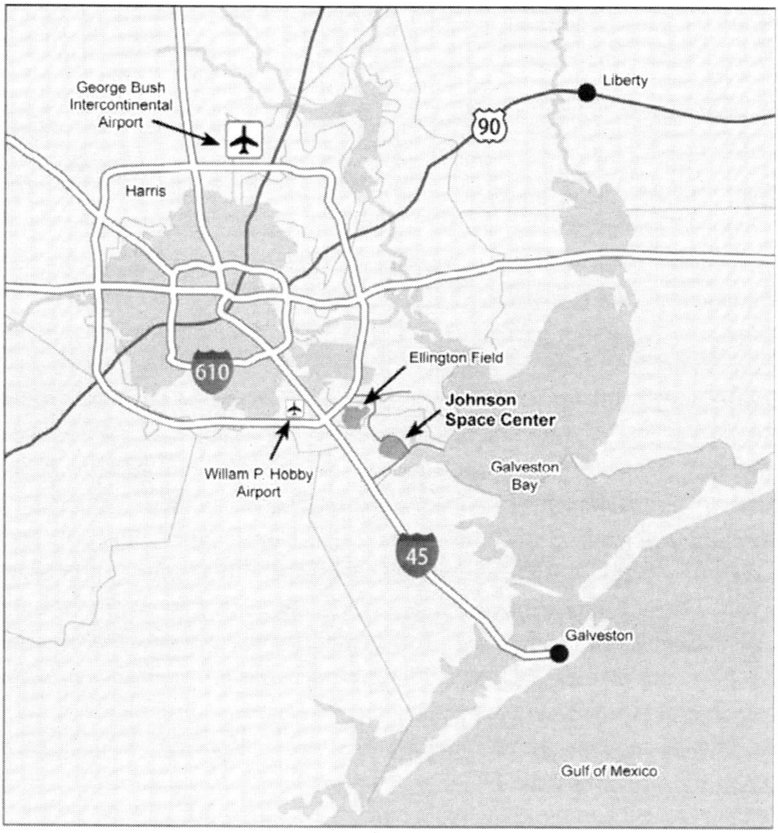

Source: NASA.

Figure 7. Location of Johnson Space Center.

Johnson Space Center's facilities are conservatively valued at $2.3 billion, and include the following:

- 163 inhabited structures;
- 4 million square feet of office space;
- 3 miles of underground tunnels;
- 8.3 miles of roadways;
- 142 labs and simulators; and
- 2 national historic landmarks, including Apollo Mission Control Center.

Among these facilities, its mission control center is often referred to as the nerve center for America's human space program. A specialized pool at the Sonny Carter Training Center near Ellington Field simulates zero gravity or weightless conditions experienced by spacecraft and crew during space flight. In addition, more than $4.0 billion of federal aerospace contracts are now managed out of Johnson Space Center, providing a local payroll of more than $1.9 billion annually. More than 15,000 people work within the center, including about 3,300 civil servants.

Climate data collected over the past 100 years in the Houston-Galveston area show a long-term pattern of relative sea level and temperature rise, according to NASA climate scientists who presented information at the March 2012 workshop. Climate models project continued relative sea level rise and warmer temperatures in the region, according to these scientists. Because of its location on the Gulf Coast, storm surge and sea level rise may be the biggest climate threats to Johnson Space Center, according to documents prepared by NASA climate scientists. Land subsidence also worsens the impacts of rising seas and storm surge. NASA climate scientists stated that, while little change is expected in average annual precipitation, precipitation could come at different intervals, and individual precipitation events may become stronger, leading to increased risks of flash flooding. In addition, according to NASA data, the number of days per year exceeding 90 degrees Fahrenheit is projected to rise dramatically in the coming century. The projected changes in the frequency of some extreme events like hot and cold days shown in table 3 would likely affect energy use and the number of hours staff can work outside.

Table 3. Changes in the Frequency of Hot and Cold Days at Johnson Space Center, as Projected by NASA Climate Scientists

Daily temperatures	Baseline	2020s	2050s	2080s
Maximum				
At or above 90°F (days)	90	100 to 113	116 to 136	126 to 164
At or above 100°F (days)	0.9	2 to 3	3 to 11	7 to 35
Minimum				
At or below 40°F (days)	37	24 to 29	17 to 26	12 to 21
At or below 32°F (days)	9	4 to 6	3 to 5	2 to 4

Source: NASA.

Note: According to a NASA document for the March 2012 Johnson Space Center workshop, these "quantitative climate projections are based on global climate model simulations conducted for the Intergovernmental Panel on Climate Change Fourth Assessment Report (2007) from the World Climate Research Programme's Coupled Model Intercomparison Project Phase 3 multi-model dataset. The simulations provide results from 16 global climate models that were run using three emissions scenarios of future greenhouse gas concentrations. The results are statistically downscaled to 1/8 degree resolution (~12 km by 12 km) based on results from the bias-corrected (to accurately reflect observed climate data) and spatially-disaggregated climate projections derived from World Climate Research Programme's Coupled Model Intercomparison Project Phase 3 data. Results provide a more refined projection for a smaller geographic area." This information is maintained at: http://gdo-dcp.ucllnl.org/downscaled_cmip3_projections and described by Maurer, E.P., L. Brekke, T. Pruitt, and P.B. Duffy (2007), 'Fine-resolution climate projections enhance regional climate change impact studies', Eos Trans. AGU, 88(47), 504.

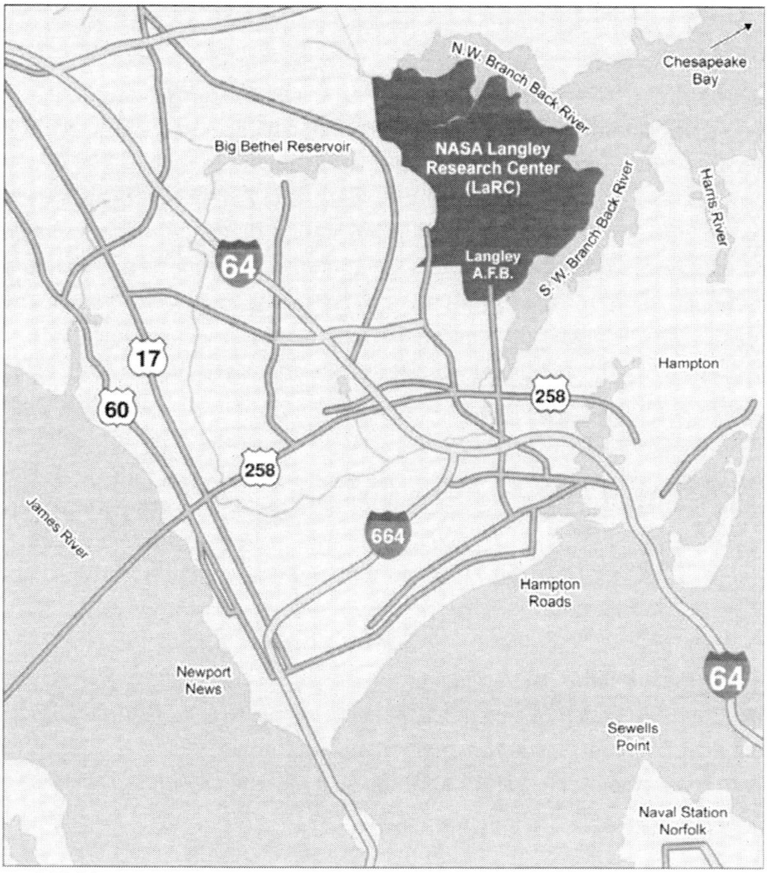

Source: NASA.

Figure 8. Location of Langley Research Center.

Langley Research Center

According to NASA documents obtained at the September 2011 workshop, Langley Research Center was founded in 1917 as the first civil aeronautical research lab, and its unique research and testing facilities make critical contributions to the development of NASA's next generation of heavy-lift rockets and capsules for future space exploration.[51] As shown in figure 8, Langley Research Center occupies nearly 800 acres in Hampton, Virginia, near the mouth of the Chesapeake Bay. The Port of Hampton Roads is the nation's third largest seaport, and the surrounding area has a strong federal presence in addition to the center, including Army, Navy, Air Force, Marines, and Coast Guard facilities. As shown in figure 8, Langley Research Center borders the Northwest Branch and Southwest Branch of the Back River, which flows east to the Chesapeake Bay. Most of its acreage is located to the west of Langley Air Force Base, with several small parcels to the east within the base.

Receiving about $927 million in federal funding in fiscal year 2012, Langley Research Center is an important part of the local economy. The center employs about 3,800 people, roughly evenly divided between civil service employees and contractors. Its 180 buildings and other facilities are valued at approximately $3.3 billion and include laboratories, aircraft simulators, and wind tunnels such as the world-unique Transonic Dynamics Tunnel used to study the effects of wind gusts on fixed-wing and rotary-wing aircraft.

According to NASA climate scientists, storm surge and relative sea level rise may be the biggest climate threats to the center. The area has always been subject to nor'easters and hurricanes, and the associated high winds and flooding. Data collected over the past 80 to 100 years in the Hampton Roads area clearly show a long-term pattern of sea level and temperature rises, accompanied by periods of shorter term variability. Climate models project continued sea level rise and warmer temperatures in the Hampton Roads region. The combination of rising sea level and severe storms could produce catastrophic impacts on the center and the surrounding assets and natural resources. Furthermore, as in the case of the Johnson Space Center, land subsidence in the area worsens the impacts of rising seas and storm surges. At the Langley Research Center workshop in September 2011, NASA climate scientists provided center staff and other local stakeholders, such as the Hampton Roads Planning District Commission, with projections of future changes in the frequency of some extreme events, such as intense precipitation, presented in table 4.

Table 4. Extreme Events at Langley Research Center: 2020-2090

Event	Direction of change	Likelihood
Heat stress	↑	Very likely
Intense precipitation events	↑	Likely
River flooding	↑	Likely
Drought	↑	More likely than not
Intense winds	↑	More likely than not

Source: NASA.

Note: This information provided by NASA is based on global climate model simulations, published literature, and expert judgment of NASA climate scientists.

DOD Facilities in Close Proximity to Langley Research Center

The climate-related challenges faced by these NASA centers are not unique to NASA and can be instructive for other types of large federal facilities. For example, while NASA facilities located on the east side of Langley Research Center are particularly vulnerable to flooding, given their low-lying coastal location, so is the adjacent Langley Air Force Base and the nearby Naval Station Norfolk, the largest naval complex in the world. The Naval Station Norfolk is home to five aircraft carriers, 70 ships, and 150 aircraft, all housed within about 7 miles of vulnerable piers and wharves located along the waterfront, according to a study under way at the U.S. Naval Academy. The relative sea level rise rate at Naval Station Norfolk is 0.2 inches per year, and the majority of the land there is less than 10 feet above mean sea level, according to this study.

The vulnerability of these nearby defense facilities can serve as a proxy for other low-lying bases worldwide and may explain DOD's focus on climate change issues. For example, DOD's 2010 Quadrennial Defense Review states that climate change and energy will play significant roles in the future security environment and that

> climate change will affect DOD in two broad ways. First, climate change will shape the operating environment, roles, and missions that we undertake…Second, DOD will need to adjust to the impacts of climate change on our facilities and military capabilities… Although the United States has significant capacity to adapt to climate change, it will pose challenges for civil society and DOD alike, particularly in light of the nation's extensive coastal infrastructure. In 2008, the National Intelligence Council judged that more than 30 U.S. military installations

> were already facing elevated levels of risk from rising sea levels. DOD's operational readiness hinges on continued access to land, air, and sea training and test space. Consequently, the Department must complete a comprehensive assessment of all installations to assess the potential impacts of climate change on its missions and adapt as required.

DOD's Strategic Environmental Research and Development Program is conducting a risk quantification study on coastal military installation assets and mission capabilities.[52] Focused on the Hampton Roads area, the objective of this study is to develop and demonstrate a multi-hazard risk assessment framework that will be suitable for evaluating changes in risks to coastal military installation assets and mission capabilities, according to DOD documents describing the study. As part of this study, DOD is developing a comprehensive inventory of assets and mission capabilities for Hampton Roads military installations, and long-term effects of increasing rates of sea level rise on the sites will be simulated over a 100-year period to estimate impacts on military installations. This information should be useful for informing NASA on the Langley Research Center and other DOD facilities.

CLIMATE CHANGE HAS NOT BEEN SYSTEMATICALLY INCORPORATED IN INFRASTRUCTURE PLANNING

Decision makers have not systematically incorporated potential climate change impacts in infrastructure planning for roads, bridges, and wastewater management systems, according to representatives we spoke with from professional associations and officials from agencies that represent or work with these decision makers. Instead, efforts to incorporate climate change impacts into planning for infrastructure projects have occurred primarily on a limited, ad hoc basis. The association representatives and agency officials told us and NRC has reported that decision makers in the infrastructure categories we examined have generally not included adaptive measures in their planning because: (1) they typically focus their attention and resources on competing, shorter-term priorities; (2) they face challenges identifying and obtaining available climate change information best suited for their projects; (3) they often do not know how to access local assistance; or (4) available climate change information does not fit neatly into their infrastructure planning processes.

Attention and Resources are Focused on Short-Term Priorities

Representatives from professional associations we spoke with said that nearer-term competing priorities make it difficult for decision makers to address the impacts of climate change, since many state and local governments responsible for the infrastructure face immediate funding and staffing challenges.[53] In many cases, according to these representatives and reports from the Transportation Research Board (TRB) of the NRC and National Drinking Water Advisory Council, adaptation is a relatively low priority compared with more traditional and immediate concerns such as managing aging infrastructure systems, sustaining current levels of service, protecting public health and safety and the environment, and maintaining service affordability.[54] In the case of wastewater infrastructure, for example, available funding is often inadequate to implement climate adaptation actions on top of more pressing needs such as meeting permit requirements, upgrading wastewater treatment plants, and preparing to implement proposed stormwater rules, according to officials from the National Association of Clean Water Agencies and a December 2010 report from the National Drinking Water Advisory Council.[55]

Due to the immediacy of many competing priorities and current funding constraints, decision makers tend to delay addressing climate change adaptation—the benefits of which may not be realized for several decades into the future.[56] As TRB reported, most infrastructure planning processes and their associated funding cycles occur on time horizons poorly matched to the longer view sometimes required to discern the effects of climate change and identify the benefits of adaptation.[57] For example, as noted in the TRB report, the longest-term planning horizons for many transportation planners rarely exceed 30 years—20 to 25 years is the norm. Yet, according to this report, the inherent variability of the climate makes it difficult to discern climate change trends over periods less than approximately 25 years. Consequently, many transportation planners perceive that the impacts of climate change will be experienced well beyond the time frame of their longest-term plans, not realizing that climate changes could already be occurring and that investment decisions made today will affect how well the infrastructure accommodates these and future changes over its design life.[58]

Decision Makers Face Challenges Obtaining the Best Available Climate- Related Information for Infrastructure Planning

Decision makers often face challenges obtaining the best available climate-related information relevant to their decision-making process.[59] According to NRC studies and decision makers and other infrastructure stakeholders we interviewed, decision makers are unsure about where to go for information and what information they should use because (1) vast amounts of information come from multiple, uncoordinated sources and (2) the quality of the information varies.

Vast Amounts of Information

Decision makers often struggle to identify which information among the vast number of climate change studies available is relevant, according to NRC studies and our interviews with federal agencies and other stakeholders. NRC researchers, federal officials, and other stakeholders reported that a vast amount of climate change information—including climate modeling results and observational datasets—is available from the independent efforts of federal and state agencies, universities, professional associations, and others. However, this information is typically made available to decision makers through what NRC described in 2012 as a "loading dock" model, which assumes that simply producing more scientific findings will improve the quality of decisions.[60] According to the NRC study, this information is reported in studies made available through peer reviewed publications and placed on the public "loading dock," where decision makers are expected to retrieve and interpret the studies for their purposes.

However, because the producers of these studies do not coordinate their efforts, the information they contain is not synthesized in ways useful to infrastructure decision makers, according to a 2010 NRC report.[61] As a result, decision makers can become overwhelmed as they attempt to piece together the information they need from the complex body of climate change literature, or can spend a great deal of time trying to find useful information. For example, one decision maker we interviewed noted that identifying the relevant aspects of the constant stream of scientific papers he receives is akin to "picking needles out of the hay." According to the 2010 NRC report, the end result of this information not being easily accessible is that people may make decisions—or choose not to act—without it.

Varied Information Quality

Given the large volume of climate-related information, decision makers also struggle to identify which information is of the best quality.[62] In many instances, according to the 2012 NRC report on climate models, decision makers often do not have sufficient information to appreciate the strengths and weaknesses of different information because differences and uncertainties among datasets and their usefulness for different purposes may not be documented. As a result, decision makers must assess the quality of information themselves and figure out how to appropriately and reliably use the results.[63] According to one representative from the Georgetown Climate Center, decision makers at the local level may be interested in incorporating climate change into their planning and design decisions but are nervous to do so because they do not know how to assess the quality of the information.

Decision Makers Face Difficulty Accessing Local Assistance

Decision makers face difficulty accessing local assistance as they consider adaption options. According to a 2010 NRC study, no one-size- fits-all adaptation option exists for a particular climate impact because climate change vulnerabilities can vary significantly by infrastructure category, region, community, or institution.[64] In other words, all adaptation is local.[65] Decision makers—who, in this case, specialize in infrastructure planning, not climate science—need assistance from experts who can help them translate available climate change information into something that is locally relevant. However, decision makers face difficulty accessing such local assistance because (1) individuals qualified to translate science to decision makers are in short supply and (2) when qualified translators do exist, decision makers do not know how to find them.

Climate information translators are in short supply. As NRC reported in 2010, a limited number of people are qualified to communicate science in ways that are useful to decision makers who are considering options for climate change adaptation.[66] Decision makers need to work with an individual who has knowledge of the present state of climate science and ability to access climate data, interpret them in a local context, and help them understand the implications of those data and attendant uncertainties, according to a 2012 NRC study on climate models.[67] As more and more communities become aware of the potential need for adaptation, intermediaries who can help bridge the gap between decision makers who want to use climate change information and the scientists who produce it are increasingly in demand. However, according to a 2011 NOAA report, meeting this increased demand presents challenges because academic institutions do not typically recognize "use-inspired" knowledge developed in collaboration with practitioners and decision makers as activities meeting academic standards for tenure, which may discourage researchers from developing such expertise.[68] In addition, some of the stakeholders we interviewed noted that, while they saw a local demand for outreach efforts to bridge the communication gap between decision makers and climate scientists, few federal programs are designed to support such activities.

Decision makers do not know where to find climate information translators. Decision makers face a challenge finding experts who can help them understand and use available climate change information. Several stakeholders we interviewed told us that federal science agencies are not in tune with the information needs of different sectors, and the disparate sources of expertise leave users confused about where to turn for help. As stated by a May 2012 NOAA-sponsored study, for most decision makers "it is not obvious who to contact for what they need, be it data, information, models or technical assistance."[69]

Available Climate-Related Information Does Not Fit Neatly within Existing Infrastructure Planning Processes

Even where good scientific information is available, it may not be in the actionable, practical form needed for decision makers to use in planning and designing infrastructure. Such decision makers work with traditional engineering processes, which often require very specific and discrete information, but scientists commonly produce climate-related information without these explicit needs in mind. Consequently, according to professional association representatives, decision makers often do not have "actionable science" of the type and scale they need to make infrastructure decisions. Specifically, (1) infrastructure decision makers need climate information at a regional or local geographic scale, but climate information has generally been produced at a global or continental scale; (2) infrastructure design decisions are made using data on the frequency and severity of extreme events, but climate information is typically presented as changes in average conditions; and (3) traditional engineering practices rely on using backward-looking historical data, whereas climate change projections are inherently forward-looking and uncertain.

Information mismatch in geographic scale. As reported by NRC in 2009, the geographic scale at which climate change information is typically available can present serious challenges for its usefulness to decision makers.[70] In general, climate change projections have focused on the global or continental scale, but the vast majority of infrastructure decision makers require information at the regional or local scale. For example, a bridge designer may require information about how climate change will impact the flow of a specific river that a bridge crosses. To generate such information at the required scale, various "downscaling" methods exist. However, these methods introduce an additional level of uncertainty, and "downscaled" information is not available for all locations because of modeling resource constraints.[71]

Climate averages versus extremes. Climate change projections tend to focus on average changes in climate variables, such as temperature and precipitation, and are not sophisticated enough to adequately characterize extreme events, which drive the design criteria for infrastructure, according to studies we reviewed and stakeholders we interviewed. Representatives of the American Society of Civil Engineers told us that climate and weather modeling indicate that extremes may become more frequent or severe, but that such modeling does not make this information sufficiently quantitative to serve as the basis for design, operation, and maintenance decisions. According to these engineers, information on future extreme events expected to occur during the service life of infrastructure is a critical component in designing more resilient infrastructure. However, according to technical comments from CEQ, the Office of Science and Technology Policy (OSTP), and USGCRP, although knowing the magnitude of future extremes would be useful, it is not necessary, for example, to know exactly how extreme precipitation will be in the future to know that larger culverts need to be used than were used in past road design.

Forward- versus backward-looking. Climate change projections are inherently forward-looking and uncertain, but traditional engineering processes rely on historical information.[72] In addition, as reported by NRC in 2012, such climate change projections commonly provide a range of possible future outcomes.[73] For example, available information may indicate that, in a particular area, intense downpours will become more frequent over the coming decades and provide a range of possibilities for the timing and magnitude of the increase. However, as stated by representatives of the American Society of Civil Engineers that we interviewed, existing infrastructure planning processes, and the design standards they rely on, require climate data with known and static probability distributions, such as the magnitude of a 100-year storm as determined by a historical record of precipitation.[74] In fact, engineers use statistical tables of historical precipitation intensity, duration, and frequency developed by NOAA that, in some cases, have not been updated since the 1960s.[75]

In light of these issues, according to the American Society of Civil Engineer representatives, climate change projections are a long way from being translatable into engineering standards of practice. As a result, NRC, in 2010, reported that adapting the nation's infrastructure to climate change will require new approaches to engineering analysis, such as using risk management to take uncertainties into account.[76] In technical comments, CEQ, OSTP, and USGCRP noted that this may overstate the issue because even historical data contain uncertainty in the timing and intensity of events, and engineering processes already account for other factors that are projected with uncertainty such as changing development patterns and population growth.

Key Factors Enabled Some Decision Makers to Integrate Climate Change into Infrastructure Project Planning

Some Decision Makers Integrated Climate Change into Project Planning Notwithstanding the challenges that have deterred most decision makers from integrating climate change considerations into infrastructure planning processes, we identified and visited several locations where some decision makers overcame these challenges. Key factors enabled these decision makers to successfully integrate climate change into their infrastructure project planning.

Some Decision Makers Integrated Climate Change into Project Planning

Decision makers at the seven locations we visited were able to integrate climate-related information into infrastructure project planning to varying degrees. These locations exhibited considerable diversity in the types of infrastructure at issue, geographic settings, and other circumstances. The adaptive measures themselves did not involve major overhauls of project plans or infrastructure systems but instead provide examples of practical responses to observed or projected climate-related impacts. Decisions to adapt infrastructure to climate change may depend on its remaining useful life, among other factors, because adaptation can be relatively more expensive when undertaken retroactively than at the design phase of a project. It is important to note that climate change was not always the primary reason for changing the infrastructure projects in these examples. Rather, the examples illustrate a shift in thinking where climate change is considered one of many hazards accounted for in planning and implementation.

Interstate-10 Twin Span Bridge (Louisiana)

As discussed above, the Interstate-10 Twin Span Bridge, which crosses Lake Pontchartrain outside New Orleans, Louisiana, is vulnerable to storm surge caused by hurricanes. Following failure of the old bridge during Hurricane Katrina, Louisiana state transportation officials decided to raise and strengthen the new Twin Span Bridge to protect against future storms—specifically to protect the structure against storm surges of similar strength to Hurricane Katrina, the largest storm surge on record for Lake Pontchartrain. When deciding how to manage risk over the bridge's intended 100-year life span, the Twin Span's design team considered many factors, such as durability, cost, and long-term maintenance. The design team ultimately decided to make a larger initial investment and build a stronger bridge to minimize future maintenance problems and expenses. The new bridge cost more than $700 million and was fully funded by federal emergency relief funds.[77]

Decision makers integrated several adaptive measures into the new bridge's design. As shown in figure 9, these measures included the following:

- *Opening railings* to reduce wave forces on the bridge's deck.
- *Raising piers* above historic peak wave heights, which involved raising the new bridge 23 feet above the old bridge elevation.
- *Lengthening piles*, long columns driven deep into the soil to support the bridge, to accommodate larger anticipated wave loads.
- *Introducing rigid connections* made of formed concrete to prevent the deck from floating off bridge piers, which occurred during Hurricane Katrina.
- *Strengthening bridge-supporting girders* with higher density high-performance concrete. This is expected to increase the bridge's

resilience to saltwater in Lake Pontchartrain, according to Louisiana state transportation officials.

According to officials from the Louisiana Department of Transportation and Development, these adaptive measures performed well during Hurricane Isaac in 2012, the first major storm to hit the new bridge since it opened to the public. When we visited after Hurricane Isaac, there were few visible impacts on the bridge structure. Although the storm surge from Isaac submerged the approaches to the bridge (i.e., the part of the bridge that carries traffic from land to the main parts of the bridge) and eroded adjacent land, the storm's impact on the bridge itself was limited to damaged signage and electrical components. Louisiana transportation officials noted that the new Twin Span's resilience during Isaac highlights the importance of designing resilient long-lived infrastructure.

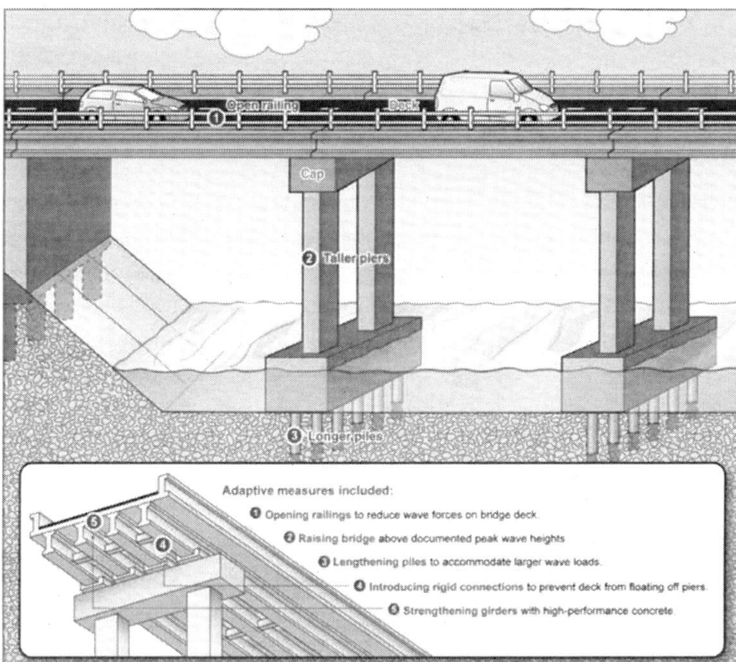

Source: GAO.

Figure 9. Adaptive Measures Integrated into New Twin Span Bridge.

Source: LA 1 Coalition.

Figure 10. Rendering of Raised Segment of Louisiana State Highway 1.

Louisiana State Highway 1

Louisiana State Highway 1 is vulnerable to storm surge given sea level rise, land subsidence, and its close proximity to the open water and the Gulf of Mexico, as previously explained. A coalition of state and local officials worked together to obtain funding to raise an 11-mile segment of the highway by 22.5 feet to protect the road from 100-year flood events. To further protect the road from storm surge, bridge designers used restraining devices and anchor bolts to prevent the road deck from dislodging from the rest of the structure in the event of a large storm surge. Figure 10 presents a rendering of the new, raised road that was opened to traffic in 2009 (on the left) in relation to the old, unraised road (on the right).

The raised segment of Louisiana State Highway 1 was largely unaffected by Hurricane Isaac—the first major hurricane to hit since the raised segment was open to the public. Some signs were damaged, but the raised section's superstructure, which includes the girders, was unaffected despite the approximately 6.5-foot storm surge measured at Port Fourchon, according to local transportation and port officials we spoke to during an on-site, follow-up visit. In contrast, the unraised sections of the highway both north and south of the raised road were damaged.[78] Figure 11 documents Hurricane Isaac-related flooding on the unraised section of Louisiana State Highway 1 north of the raised road.[79]

Washington State Route 522

Washington State Route 522 and its Snohomish River Bridge are vulnerable to projected increases in precipitation and flash flooding, which may lead to increased bridge scour and roadbed damage. In 2008, the Washington State Department of Transportation completed environmental reviews for a major construction project along Route 522 to improve safety and reduce congestion. During the design, state officials integrated several measures in the project that both reduced the project's impact on the environment and increased its resilience to projected climate change impacts.

Figure 12 illustrates some of the measures integrated into the project design. Specifically, at the Snohomish River Bridge site, engineers deepened bridge footings—the enlarged portions of bridge foundations that rest directly on soil, bedrock, or piles—to protect against the effects of changes in the flow of the river. Engineers also placed bridge piers at least 10 feet above documented peak flows and aligned the bridge at the least vulnerable location along the river. Furthermore, state transportation officials built five stormwater treatment areas and eight water retention ponds that will serve the dual purposes of controlling and treating storm water flows, and plan to increase the size of two drainage culverts, to (1) mitigate the project's impact on the surrounding environment by allowing wildlife to cross between habitat areas and improving fish access; (2) protect the roadbed by allowing greater amounts of water to flow more freely, preventing damaging roadbed saturation; and (3) increase the connectivity of waterways, which preserve natural drainage.

Source: LA 1 Coalition.

Figure 11. Flooding of an Unraised Segment of Louisiana State Highway 1 Following Hurricane Isaac.

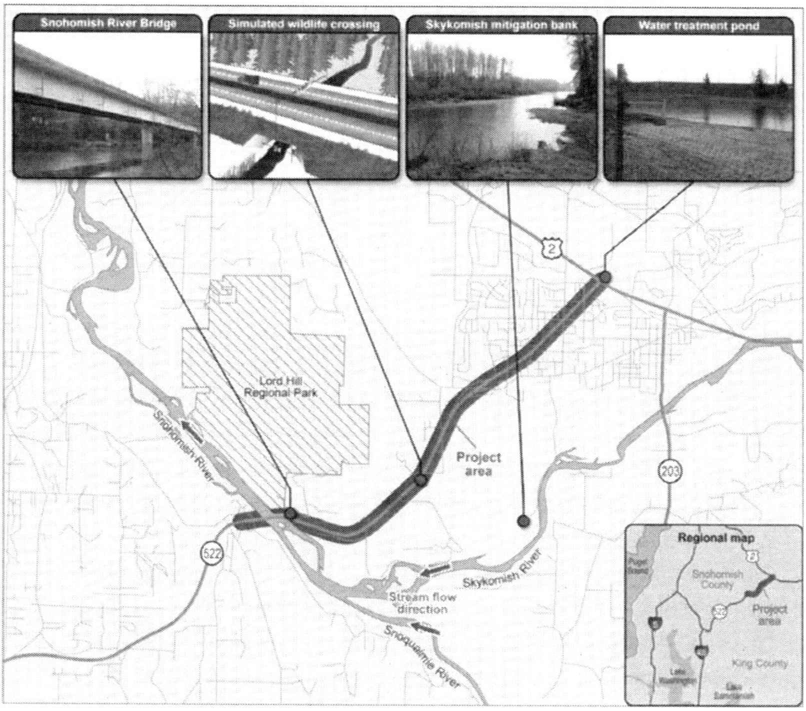

Source: Parametrix (map); Washington State DOT (wildlife crossing diagram); and GAO (photos).

Figure 12. Adaptive Measures Integrated into Washington State Route 522.

Also shown in figure 12, is the Skykomish Basin wetland mitigation bank[80] located upstream of the Snohomish River Bridge. For this project, purchasing credits from the mitigation bank serves the dual purposes of (1) offsetting the loss of 15.6 acres of wetland and wetland buffer areas damaged during construction with compensatory flood storage and (2) reducing the erosive capacity of water on the bridge by slowing the flow of the Skykomish River.

King County, Washington, Wastewater Treatment Division

Facilities managed by the King County Wastewater Treatment Division are vulnerable to sea level rise, which may increase flooding of infrastructure and combined sewer overflows. To address this concern, the Wastewater Treatment Division made minor modifications to new construction and rehabilitation projects and plans to more formally incorporate climate change information into its asset management program. Based on a climate change vulnerability assessment of its system, engineers adjusted the design of two vulnerable facilities. First, engineers determined that raising the new Brightwater Flow Meter Vault and Sampling Facility's equipment by 5 feet would address these assets' vulnerabilities to projected sea level rise. Accordingly, these facilities were designed and built 5 feet higher. Second, at the Barton Pump Station, which was scheduled for rehabilitation, engineers raised an overflow weir and installed a flap gate, pictured in figure 13, to prevent saltwater intrusion.[81]

According to King County Wastewater Treatment Division officials, these adaptive actions were "low-risk, high-reward" measures, illustrating "no regrets" solutions that provide benefits regardless of future climate conditions. For example, the modifications made to the Barton Pump Station will help protect against current saltwater intrusion problems such as the event that tripped off a combined sewer overflow alarm in January 2010 during a particularly high tide.

Milwaukee Metropolitan Sewerage District (Wisconsin)

Milwaukee Metropolitan Sewerage District facilities are vulnerable to projected increases in frequency and intensity of extreme rainfall events due to climate change, potentially resulting in more frequent and larger combined sewer overflows. As part of broader efforts to meet growing demand for sewer capacity, Milwaukee Metropolitan Sewerage District officials employed what they called "green infrastructure" programs to make the district's sewer system more resilient to climate change by capturing and holding or slowing the flow of stormwater, and officials plan to incorporate climate change adaptation into infrastructure planning and design where it makes sense as their facilities age and are replaced over time. Three of these programs, shown in figure 14, include (1) bio-swales, which are depressed catchment areas planted with vegetation to capture and infiltrate stormwater runoff; (2) green roofs either partially or completely planted with vegetation to hold rainwater; and (3) the purchase of undeveloped property to preserve targeted land areas to store and drain stormwater runoff into the ground naturally.[82]

Milwaukee Metropolitan Sewerage District officials emphasized the co-benefits of green infrastructure programs, including flood management, improved air and water quality, increased property values, reduction of urban heat island effect, and additional recreational amenities.

Source: King County Wastewater Treatment Division.

Figure 13. Flap Gate Installed at the Barton Pump Station.

Source: Milwaukee Metropolitan Sewerage District.

Figure 14. Examples of Green Infrastructure Projects in Milwaukee.

NASA Johnson Space Center (Houston, TX) and NASA Langley Research Center (Hampton, VA)

Storm surge and relative sea level rise pose significant climate threats to Johnson Space Center and Langley Research Center. As previously discussed, these centers hosted adaptation workshops to identify risks to assets and capabilities from current and future changes in the climate. We attended these workshops and observed that they involved a broad range of stakeholders—including NASA climate scientists, headquarters officials, and center staff; local government and industry officials; and experts from local academic institutions—in a comprehensive evaluation of center vulnerability. The workshops are organized to help each center (1) obtain information on historic, current, and projected climate hazards specific to the region; (2) characterize the risk of current and future climate on center systems, assets, and capabilities; (3) start to build capacity to execute a continuous adaptation process; and (4) begin to plan for the future and integrate climate considerations into existing management plans and processes. These workshops were held in late 2011 and 2012, so it is too early to fully evaluate the progress of NASA centers in incorporating climate change into their planning processes.

NASA officials have begun to conduct follow-up activities and analyze lessons learned from the workshops. An important outcome of the workshops has been increasing NASA collaboration and partnership with surrounding communities, federal neighbors, and academia, according to NASA officials. Additionally, some centers are supporting local tidal marsh restoration projects or implementing their own protective measures of vulnerable mission-critical areas. Low impact development has been implemented as one way of slowing water runoff and allowing more infiltration. For example, the Langley Research Center has identified high- priority areas for wetland development to act as buffer zones for future storm surge events, and it planned to harden or elevate vulnerable infrastructure elements (heating, ventilation, and air-conditioning, as well as electrical transformers) as it rehabilitates, repairs, and maintains its assets over time.

Key Factors Enabled Decision Makers to Consider Climate Change in Project Planning

The adaptive measures described above did not necessarily require decision makers to undertake major changes to project plans or infrastructure systems but often did involve a commitment of financial resources and, importantly, a change in mind-set toward addressing longer-term and uncertain risks that many decision makers are not yet in a position to consider. Key factors that enabled these decision makers to undertake such measures and overcome the challenges that have deterred others from integrating climate change into infrastructure planning were that (1) their local circumstances were conducive to addressing climate-related risks, (2) they learned to use available climate information, (3) they had access to local assistance, or (4) they considered climate impacts within existing planning processes in the same context as other potential risks.

Local Circumstances Were Conducive to Addressing Climate-Related Risks

At the sites we visited, local circumstances were conducive to addressing climate-related risks because these sites: (1) were in regions that recently experienced a natural disaster or that had discernible climate-related impacts, providing a stronger basis for engaging in adaptation efforts; (2) had strong community leadership to help spur action; and (3) had executive orders or other formal policy documents to help justify and encourage taking adaptive actions.

Recent extreme weather events triggered a response. In some cases, decision makers were compelled to account for future climate conditions by a triggering event that demanded a response or created a policy window for action. For example, Hurricane Katrina exacted a heavy toll on the old Twin Span Bridge, necessitating a rebuild and prioritizing the construction of a new, more resilient bridge. As noted in the 2009 NRC report on climate-related decision support, recent firsthand experience with a natural disaster, such as a heat wave, drought, storm, or flood, can dramatically increase decision makers' desire for, and openness to, new information and action.[83] For example, according to stakeholders from the American Association of State Highway and Transportation Officials (AASHTO), the sense of urgency of climate change adaptation is generally higher in coastal states and in areas that have experienced recent events affecting their transportation infrastructure. Similarly, EPA officials told us that the likelihood that a wastewater utility would consider climate change in infrastructure planning depends largely on, among other things, where it was located geographically and, in some cases, whether it had already experienced a weather event that might increase with a changing climate. This point was evident during our visit to Milwaukee, where extreme rainfall events in 2008, 2009, and 2010 each exceeded the magnitude of a 100-year storm, making the public aware of the need to prepare for the impacts of climate change.[84] Also, according to NASA officials, the impact of extreme events on the two NASA centers we visited helped drive the creation of the adaptation workshops.

Strong leadership spurred action. In other cases, strong leadership compelled decision makers to account for future climate conditions. A 2010

NRC report notes that the engagement of and direct input from a chief executive is typically required to make the case that adapting to anticipated climate change is important.[85] Further, according to a 2009 NRC report, the leadership of top-level individuals is often necessary to overcome deeply engrained barriers.[86] According to officials from the Milwaukee Metropolitan Sewerage District, the district's Executive Director championed climate change adaptation, and his efforts, along with support from the district's commission, have been critical to the district's success in upgrading its sewer system. In addition, senior leadership at NASA headquarters and the two NASA centers we visited clearly called for the consideration of climate risks and adaptation strategies in infrastructure plans and processes enabling some NASA centers, including Johnson Space Center and Langley Research Center, to begin to do so. At the Langley Research Center workshop in September 2011, the Director of Center Operations notified his supervisory staff that they would have to schedule a briefing with the Director outlining how they will start to incorporate consideration of climate risks and adaptation strategies into their plans and processes.

Policy documents helped justify action. As shown by our site visits, executive orders or other formal policy documents can help justify and encourage adaptive efforts at the state and federal levels.[87] For example, Washington State Executive Order 07-02, issued in 2007, directed the development of a climate change initiative to determine the specific steps that should be taken to prepare for the impact of global warming on infrastructure, among other things. Since then, state transportation officials considered climate change adaptation during the environmental review of Washington State Route 522, and the Washington State Department of Transportation has directed all project teams to consider climate change in their national and state environmental review documents. Similarly, King County's 2007 Climate Action Plan provided the impetus to move forward on adaptation activities, according to Wastewater Treatment Division officials. At the federal level, the October 5, 2009, Executive Order 13514 on Federal Leadership in Environmental, Energy, and Economic Performance directs federal agencies to evaluate their climate change risks and vulnerabilities and manage the effects of climate change on the agency's operations and mission in both the short- and long-term. NASA officials at the Johnson Space Center and Langley Research Center workshops cited the executive order as a reason to take the workshops seriously.

Table 5. NASA Model Results of Projected Climate Changes at Johnson Space Center

Variable	Baseline	2020s	2050s	2080s
Average temperature	70.5°F	+1.5 to 2.5°F	+2.5 to 4.5°F	+3.5 to 7°F
Annual precipitation	54 inches	-5 to +5 %	-15 to +5 %	-15 to +5 %
Sea level rise	Not applicable	+2 to 3 inches	+ 5 to 9 inches	+11 to 20 inches

Source: NASA.

Notes: According to NASA's workshop materials for Johnson Space Center, temperature and precipitation projections reflect a 30-year average centered on the specified decade; sea levels are averages for the specific decade. The baseline for temperature and precipitation is the most complete 30-year data period centered around the 1980s; the baseline for sea level is 2000-2004. Temperature and precipitation data are for Houston, TX (Hobby Airport), and sea level data are for Galveston, TX. Temperatures are rounded to the nearest half degree, precipitation projections to the nearest 5 percent, and sea level rise to the nearest inch. Shown is the central range (middle 67 percent of values from model-based probabilities) across the general circulation models and greenhouse gas emissions scenarios. Data are from the NOAA National Climatic Data Center.

According to NASA's workshop materials for Johnson Space Center, these quantitative climate projections are based on global climate model simulations conducted for the IPCC Fourth Assessment Report (2007) from the World Climate Research Programme's Coupled Model Intercomparison Project Phase 3 multi-model dataset. The simulations provide results from 16 global climate models that were run using three emissions scenarios of future greenhouse gas concentrations. The results are statistically downscaled to 1/8 degree resolution (~12 km by 12 km) based on results from the bias-corrected (to accurately reflect observed climate data) and spatially- disaggregated climate projections derived from World Climate Research Programme's Coupled Model Intercomparison Project Phase 3 data. Results provide a more refined projection for a smaller geographic area. This information is maintained at: http://gdo-dcp.ucllnl.org/downscaled_cmip3_projections and described by Maurer, E.P., L. Brekke, T. Pruitt, and P.B. Duffy (2007), 'Fine-resolution climate projections enhance regional climate change impact studies', Eos Trans. AGU, 88(47), 504.

Decision Makers Learned to Use Available Information

The examples from our site visits show that it is possible to use many types of climate-related data to make more informed decisions about climate change in project-level infrastructure planning. Importantly, the decision makers at the sites we visited did not wait for perfect information to take action, and they learned to manage the uncertainty associated with climate-related data. As stated to us by an official from Seattle Public Utilities, "uncertainty should not be an excuse for inaction on climate change adaptation. Decision makers have to get smarter and find ways to incorporate whatever climate information they have."[88] Despite the challenges that decision makers reported in identifying and applying available information about climate change, decision makers at the sites we visited learned to use a range of information sources, including (1) site-specific projections of future climate conditions, (2) qualitative information based on state or regional scale climate projections, and (3) observed climate data.

Site-specific projections of future climate conditions. In some cases, decision makers learned to use site-specific projections of future climate conditions when determining how to take adaptive measures. For example, NASA climate scientists prepared downscaled climate variable projections for the Johnson Space Center and Langley Research Center workshops. Table 5 shows projected quantitative climate changes for Johnson Space Center.

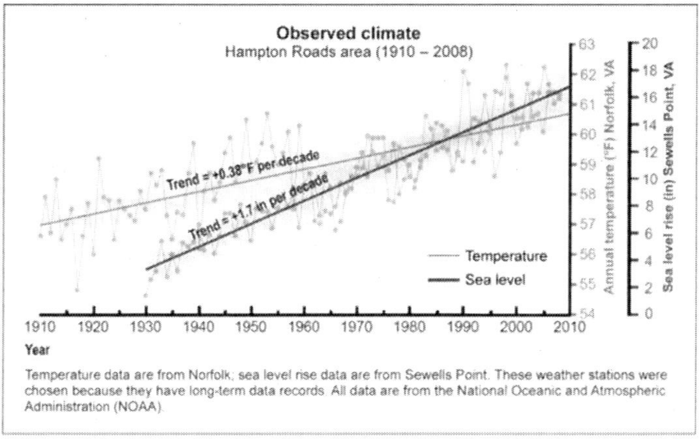

Source: NASA.

Note: NOAA's weather forecast offices maintain a network of thousands of weather-monitoring stations throughout the United States, some with temperature records stretching back well over a century. These weather-monitoring stations are equipped to measure meteorological conditions at ground level—usually the daily maximum and minimum temperatures and 24-hour precipitation totals—to support weather forecasts and for the study of climate.

Figure 15. NASA Model Results of Observed Climate Changes at Langley Research Center.

Furthermore, Milwaukee Metropolitan Sewerage District officials used site-specific climate change projections prepared by the Wisconsin Initiative on Climate Change Impacts as a foundation for planning green infrastructure components that may have a beneficial impact to their system. More specifically, the Milwaukee Metropolitan Sewerage District contracted with researchers at a local academic institution to use these projections to provide an analysis of how climate change could impact the sewer system and cause sewer overflows. The King County Wastewater Treatment Division similarly used sea level rise projections from the University of Washington's Climate Impacts Group in its facilities vulnerability study.

Qualitative information. Not all decision makers have access to quantified site-specific projections of future climate changes. In the absence of such projections, some infrastructure decision makers from our site visits used qualitative evaluations of state or regional scale climate projections to help make more informed decisions. For example, site-specific climate projection data were not available when Washington State Department of Transportation officials evaluated adaptation measures for Washington State Route 522. For this reason, the project team conducted a qualitative evaluation of climate variability based on available information, such as information from the region's transportation planning organization and studies reflecting how climate change impacts may manifest themselves within the region. Similarly, when Seattle Public Utilities officials assessed their adaptation options, site-specific climate change projection data were not adequate to be useful for planning purposes. As a result, according to a 2011 EPA report, utility officials used their general understanding of climate trends to apply a safety factor to new infrastructure construction so that that new investments would more likely perform their intended function over their useful lives.[89] This is a practical approach that can be generalized to a wide range of adaptation situations, according to technical comments from CEQ, OSTP, and USGCRP.

Observed historical climate data. According to a NOAA workshop report on climate adaptation, observed climate records help to overcome barriers that may be associated with discussions of climate change. Milwaukee Metropolitan Sewerage District officials told us they emphasize data on observed changes when the public inquires about the district's climate change adaptation actions. Similarly, officials from the Wisconsin Initiative on Climate Change Impacts stated that while it is difficult to ask a planning board for money to make design changes based on uncertain projections, observations can show that the climate is changing and that stakeholders are often more compelled by historical data than by model projections. As we observed, NASA kicked off each workshop by presenting observed climate data for the local area and discussing participants' personal experiences with weather events to make the potentia lly abstract notion of climate vulnerability "real." Figure 15 shows the observed historical sea level and temperature data that NASA used in its Langley Research Center workshop.

Some decision makers at the sites we visited said that they used historic climate data to inform engineering decisions. For example, when designing the new Twin Span Bridge, Louisiana Department of Transportation and Development engineers wanted to design the bridge to resist storm surge and wave action from the worst-case storm scenario. However, they had no detailed information about Lake Pontchartrain's wave characteristics or guidance from the AASHTO on how to design a bridge to withstand extreme weather events in coastal areas. To obtain this information, these officials hired experts in the area of wave mechanics to conduct a storm analysis. The experts used historic storm surge data to develop hypothetical scenarios regarding wave crest elevations and hurricane tracks. While reviewing historic data, the experts discovered that Lake Pontchartrain is very susceptible to storm surge. To determine the worst-case scenario for the Twin Span Bridge, they modeled a storm with properties similar to Hurricane Katrina along different storm tracks. The storm surge and waves created by a Katrina- like hurricane located west of the bridge became the basis of their design.

Decision Makers Had Access to Local Assistance

Access to local assistance was instrumental to decision makers' ability to undertake climate adaptation efforts at the sites we visited. Decision makers used this assistance to (1) translate available climate information into a meaningful and usable form and (2) help communicate to the local community the risks associated with climate change and the importance of taking action.

Translating available information. At most of our site visits, local experts helped decision makers bridge the gap between the information they needed and the science that was available. Decision makers at the sites we visited told us that local experts were instrumental because they understood the local context. In one example, the Milwaukee Metropolitan Sewerage District sought the expertise of local scientists and planners who were familiar with its sewer system and local considerations because available climate change information could not be used "off the shelf" for wastewater planning. These experts translated region-specific climate model data into a form that could be plugged into existing sewer system models used by the Milwaukee Metropolitan Sewerage District for system planning and evaluation. This enabled the district's decision makers to understand the projected impacts of climate change on its sewer system and appropriately tailor their adaptation efforts. In another example, NASA developed a Climate Adaptation Science Investigator working group with members at each of its centers to partner NASA climate scientists with local infrastructure managers, thereby developing local expertise that decision makers could use to tailor center-specific adaptation solutions.

Communicating to the public. In addition to helping translate climate change information, decision makers at our site visits noted the importance of having local experts to help communicate local climate change information to the public and help the community understand the need for adaptation. For example, several decision makers in King County said that experts from the Climate Impacts Group at the University of Washington, through outreach programs, were effective in focusing the community's attention on climate change issues and the importance of investing in climate preparedness. According to one of the decision makers, when King County officials are "able to stand shoulder-to- shoulder" with local scientists known in the community, they do not have to defend the underlying climate science to customers who could potentially face increased rates. Similarly, several decision makers in Milwaukee noted that having local experts helps the agency more effectively convey to the community the need for and importance of climate preparation. They noted, "the response you get from people when talking about climate change often depends on who is delivering the message."

Decision Makers Considered Climate Impacts within Existing Planning Processes

Some decision makers stated during our site visits that a key factor in their success was an ability to consider potential climate change impacts within their existing infrastructure planning processes so that they were viewed in the same context with other potential risks. As NRC reported in 2010, incorporating adaptation considerations into existing processes—a concept known as "mainstreaming"—can reduce costs and provide incentives to adapt.[90]

The value of mainstreaming adaptation into normal planning processes was illustrated by several of our site visits. In Milwaukee, for example, sewerage district officials noted that efforts to consider climate change in sewer infrastructure planning were successful because climate change information could be integrated into existing planning processes and analyses. In one such effort, an engineer at the Milwaukee Metropolitan Sewerage

District told us that the agency builds new water conveyance structures taller because it "makes sense" given the known vulnerabilities to increased flooding in the region. Additionally, in the Washington State Route 522 example, project planners incorporated climate change considerations during the project's environmental review process, which provided the opportunity to explain how the elements of the project helped to improve climate resiliency and reduce potential for damage from extreme storm events. According to Washington State Department of Transportation officials, climate adaptation measures were integrated with decisions about how to minimize environmental effects and comply with regulations, permits, and approvals.

Some of the decision makers from our site visits envision more formally integrating potential climate change impacts into planning processes. For example, wastewater officials from King County said that they will likely include climate change risk in a field of the county's asset management database that is maintained to track the status and condition of infrastructure components. Therefore, when a particular component is due for rehabilitation or replacement, information will be readily available for planners and designers to make the component more resilient to climate change as it is being modified anyway. Similarly, NASA's Climate Change Adaptation Policy Statement notes that the agency plans to start building the capacity to execute a continuous adaptation process and will require that climate considerations be incorporated into existing management plans and processes. According to NASA officials, such plans and processes include master planning efforts, construction of facilities projects, environmental management systems, and permitting.[91]

FUTURE FEDERAL ADAPTATION EFFORTS COULD BETTER MEET THE NEEDS OF LOCAL INFRASTRUCTURE DECISION MAKERS

Emerging federal efforts are under way to facilitate and enable more informed decisions about adaptation, including raising public awareness, but these efforts could better meet the needs of local decision makers, according to studies, decision makers from our site visit locations, and other stakeholders. In some cases, these sources identified opportunities to better meet the needs of local infrastructure decision makers in the future by: (1) improving infrastructure decision makers' access to and use of available climate-related information, (2) providing increased access to local assistance, and (3) considering climate change in existing planning processes.

Emerging Federal Efforts to Raise Public Awareness of Adaptation

Emerging federal efforts to raise public awareness of climate change adaptation include (1) the Interagency Climate Change Adaptation Task Force, (2) the National Climate Assessment status report on climate change science and impacts, and (3) vulnerability assessments for specific infrastructure categories.

Interagency Climate Change Adaptation Task Force

Executive Order 13514 on Federal Leadership in Environmental, Energy, and Economic Performance called for federal agencies to participate actively in the already existing Interagency Climate Change Adaptation Task Force.[92] The task force, which began meeting in Spring 2009, is cochaired by CEQ, NOAA, and OSTP and includes representatives from more than 20 federal agencies and executive branch offices.[93] The task force was formed to develop federal recommendations for adapting to climate change impacts both domestically and internationally and to recommend key components to include in a national strategy.

On October 14, 2010, the task force released its interagency report outlining recommendations to the President for how federal policies and programs can better prepare the United States to respond to the impacts of climate change.[94] The report recommended that the federal government implement actions to expand and strengthen the nation's capacity to better understand, prepare for, and respond to climate change. The 2010 report laid out guiding principles for adaptation for federal agencies (and that should be considered by others) and policy goals and recommended actions for the federal government. These recommended actions include making adaptation a standard part of agency planning to ensure that resources are invested wisely and services and operations remain effective in a changing climate.

On October 28, 2011, the task force released *Federal Actions for a Climate Resilient Nation: Progress Report of the Interagency Climate Change Adaptation Task Force*, which outlined federal progress in expanding and strengthening the nation's capacity to better understand, prepare for, and respond to extreme events and other climate change impacts. The report provides an update on actions in key areas of federal adaptation, including building resilience in local communities and providing accessible climate information and tools to help decision makers manage climate risks.

According to the task force, its work has increased awareness of climate change across the federal government and generated adaptive actions. In technical comments, CEQ, OSTP, and USGCRP noted that the task force recommended that each agency "mainstream" adaptation planning into its missions, operations, and facilities so as to ensure that climate change impacts are taken into consideration with long-term planning and reforming building standards. The task force also stated that, as the federal government further integrates adaptation into its operations, policies, and programs, it will catalyze additional adaptation planning across the nation. However, the 2012 NRC report on climate models describes the task force as having largely been confined to convening representatives of relevant agencies and programs for dialogue, without mechanisms for making or enforcing important decisions and priorities.[95] In technical comments, CEQ, OSTP, and USGCRP took issue with NRC's description of task force activities, citing the release of agency adaptation plans (discussed further below) and a variety of other strategic planning efforts, including the *National Fish, Wildlife and Plants Climate Adaptation Strategy*.[96]

National Climate Assessment

The National Climate Assessment, required not less frequently than every 4 years by the Global Change Research Act of 1990 and conducted under the USGCRP, analyzes the effects of global change on the natural environment, agriculture, energy production and use, land and water resources, transportation, human health and welfare, human social systems, and biological diversity, and it analyzes current trends in global change, both human-induced and natural, and projects major trends for the subsequent 25 to 100 years.[97] USGCRP intends that this assessment be used by U.S. citizens, communities, and businesses as they create plans for the nation's future.

According to USGCRP documents, these assessments serve an important function in providing the scientific underpinnings of informed policy and act as status reports about climate change science and impacts. They can identify advances in the underlying science, provide critical analysis of issues, and highlight key findings and key unknowns that can guide decision making. Assessments attempt to identify climate impacts at the regional level to raise awareness and spur more informed decision making.

There have been two assessments in the past 20 years, and a draft of a third assessment report was released for public review on January 11, 2013.[98] The first, in 2000, included a large stakeholder engagement process and the second, in 2009, was more focused on specific climate science topics. The third assessment—expected to be finalized in March 2014, according to USGCRP—differs in multiple ways from previous efforts, according to USGCRP's strategic plan.[99] Building on the recommendations of the NRC, it will both implement a long-term, consistent, and ongoing process for evaluation of climate risks and opportunities and inform decision making processes within regions and sectors. An essential component of this ongoing process is to establish a sustained assessment activity both inside and outside of the federal government that draws upon the work of stakeholders and scientists across the country. The third National Climate Assessment report will also have significant components related to transportation and water infrastructure, among other sectors, according to USGCRP.[100]

Vulnerability Assessments

Some federal agencies are also conducting vulnerability assessments for specific infrastructure categories. For example, the Federal Highway Administration is developing a vulnerability and risk assessment model for transportation infrastructure. To test this effort, the Federal Highway Administration funded pilot studies in Washington State; the San Francisco Bay Area; Oahu, Hawaii; Hampton Roads, Virginia; and New Jersey.[101] For these pilots, the Federal Highway Administration developed a risk assessment model to aid state departments of transportation and metropolitan planning organizations in inventorying assets, gathering climate information, and assessing the risk to their assets and the transportation system from climate change.[102] The pilots started at the end of 2010 and participating agencies completed their project reports in late 2011. According to agency officials, the Federal Highway Administration is initiating a second round of pilots, to be launched in early 2013, with an expanded focus on extreme weather events and adaptation options.

The Federal Highway Administration has reviewed these reports and used the feedback from the pilot agencies to refine the vulnerability and risk assessment framework, according to agency officials. Specifically, the Federal Highway Administration's December 2012 *Climate Change and Extreme Weather Vulnerability Assessment Framework* draws from the experiences of these pilot projects to develop a guide for transportation agencies interested in assessing their vulnerability to climate change and extreme weather events.[103]

The framework gives an overview of key steps in conducting vulnerability assessments and uses examples to demonstrate a variety of ways to gather and process climate-related information. Federal Highway Administration officials also noted that the agency is currently soliciting proposals for additional pilot agencies to further evaluate the framework.[104]

Improving Infrastructure Decision Makers' Access to and Use of Available Information

According to relevant studies, local decision makers from our site visits, and other stakeholders, future federal efforts to improve access and use of available climate-related information could better focus on the needs of local decision makers. These sources identified opportunities for these efforts to better meet the needs of local infrastructure decision makers in the future by (1) better coordinating and improving access to the best available climate-related data and (2) providing technical assistance to help local decision makers translate available climate-related data into information useful for decision making.

Coordinating and Improving Access to the Best Available Climate-Related Data

Emerging federal efforts to coordinate and improve access to the best available climate-related data for decision making are much needed, according to studies, local decision makers from our site visits, and other stakeholders. According to a 2010 NRC study, the federal government has a critically important role in coordinating available climate-related data because it provides and supports large infrastructure for data collection and analysis (e.g., satellites, climate models, and monitoring systems), and can set standards for information quality.[105] However, as noted by USGCRP in its April 2012 strategic plan, federal agencies generally have pursued a distributed data strategy over the last decade, in which individual agencies have established archives for collecting and storing data.[106] This means that decisions and actions related to climate change are currently being informed by a loose confederation of networks and other institutions, according to the 2010 NRC study.

A range of stakeholders cited the need to improve the coordination of agency climate data collection and consolidation efforts. For example, Milwaukee Metropolitan Sewerage District officials told us they believe the federal government could better focus its initiatives by integrating climate-related information programs under one umbrella. Echoing this sentiment, officials from the Wisconsin Initiative on Climate Change Impacts stated that the "federal agencies that provide climate change information need to find a way to coordinate their efforts. Currently, there is no coherence among such agencies." In addition, in its December 2010 report, the EPA National Drinking Water Advisory Council noted that there is a pressing need for a coordinated, collaborative, information strategy that is supported by the key agencies and organizations and that helps make the most effective use of limited financial and technical resources available to address climate change challenges.[107]

Federal entities are beginning to respond to the need for improved coordination of agency activities to organize available climate data. For example, USGCRP's April 2012 strategic plan recognizes that for the past two decades:

> USGCRP agencies have been providing global change information that is essential to many aspects of policy, planning, and decision making. The growing demands for information by decision makers, however, are highlighting the need for improved accessibility to more comprehensive, consolidated, and user- relevant global change-related data and information. Global change observations, monitoring, modeling, predictions, and projections—underpinned by the best-available natural and social science—can provide the framework of global change information. No single agency can provide the breadth of information needed. This provides a unique opportunity for current and potential USGCRP partners, including the private sector, academia, and other Federal agencies, to improve the effectiveness of its global change information in ways that better address the growing public demand for science that can inform decision making without prescribing outcomes.[108]

USGCRP has established an adaptation science workgroup focused on coordinating interdisciplinary science in support of national and regional adaptation decisions, among other activities, and is working with CEQ, OSTP, and other agencies to improve coordination of the development and delivery of climate science and services to local decision makers, according to USGCRP officials.

In our 2011 report on climate change funding, OSTP stated that, while significant progress is being made in linking the climate science-related efforts, individual agencies still want to advance initiatives that promote or serve their agency missions.[109] This, according to OSTP, yields a broader challenge of tying climate-related efforts together into a coherent governmentwide strategy since interagency coordinating programs like USGCRP generally do not have direct control over agency budgets. According to a 2009 NRC report, the absence of centralized budget authority limits the ability of the USGCRP to influence the priorities of participating agencies or implement new research directions that fall outside or across agency missions.[110]

In technical comments, CEQ, OSTP, and USGCRP noted that the absence of centralized budget authority remains the most important impediment to USGCRP's ability to meet its mandate to provide the information needed to support adaptation planning and implementation. However, according to the technical comments, agencies' enabling legislation and subsequent reauthorizations generally require that they advance initiatives that promote or serve their agency missions, and the appropriations process supports and reinforces separate budget authorities, particularly where agencies are covered by different Congressional committees. The technical comments also noted the difficulty in finding mechanisms to facilitate joint federal funding of projects makes collaboration and implementation of joint priorities more challenging.

While coordinating available climate-related data is a first step in making more informed adaptation decisions, another key step is to ensure decision makers have access to the best available data. According to a 2010 NRC study, an informed and effective national response to climate change requires that the widest possible range of decisions makers— public and private, national and local—have access to up-to-date and reliable information about current and future climate change, the impacts of such changes, the vulnerability to these changes, and the response strategies for reducing emissions and implementing adaptation.[111] As stated by AASHTO officials, the most important role that the federal government could play in the transportation sector with respect to adaptation would be to provide a central repository for state transportation officials to go to for data. Similarly, stakeholders at a recent NOAA-sponsored workshop on transportation infrastructure adaptation highlighted the importance of clear guidance on where to look for information, including the need for a central clearinghouse for climate and weather information relevant to transportation officials.

Efforts to provide infrastructure decision makers with access to climate-related information are an emerging priority across the federal government. For example, on June 6, 2012, both the Acting Director of OMB and the Director of OSTP signed the Science and Technology Priorities for the Fiscal Year 2014 Budget memorandum, which states that agencies should give priority to research and development that strengthens the scientific basis for decision making. Such research and development is to include efforts to enhance the accessibility and usefulness of data and tools for decision support, specifically efforts that advance the implementation of federal adaptation initiatives. USGCRP's April 2012 strategic plan recognizes this high-level priority by identifying enhanced information management and sharing as a key objective.[112] In this regard, USGCRP is pursuing the development of a Global Change Information System to support coordinated use and application of federal climate science.[113] USGCRP plans to leverage existing tools, services, and portals from the USGCRP agencies to develop a "one-stop shop" for accessing global change data and information, according to the strategic plan.

These efforts, if fully implemented, appear likely to improve access to the broad range of available climate-related information. However, it remains unclear how federal efforts will address the challenge of clearly identifying the best available information to use in local infrastructure planning so decision makers who may not be familiar with climate science are not left to sort it out themselves.

Several site visit decision makers, infrastructure stakeholders, and available studies noted additional infrastructure adaptation information needs that could be met through future federal research. Better organized and accessible climate data may meet some of these needs, but a "one- stop-shop" may also highlight gaps in existing data. In other words, access to existing information may not be enough to meet all the perceived needs of infrastructure decision makers because some types of desired information do not yet exist. According to OMB's and OSTP's fiscal year 2014 science and technology priorities memo, specific areas where progress is needed include: observations to detect trends in weather extremes; integration of observation into models; simulation and prediction at spatial and temporal scales conducive to decision making; and adaptation responses to changing frequency and intensity of extreme weather events. Regardless, improved coordination and consolidation of federal climate data will assist in the prioritization of future federal adaptation science activities and help local and federal officials clarify true "needs" from "wants."[114]

Providing Technical Assistance to Help Interpret Climate- Related Data

Even with coordinated and accessible climate data, local decision makers will need technical assistance and tools to interpret what the data mean for infrastructure planning, according to our 2009 report on climate change adaptation.[115] For example, for its workshops at Johnson Space Center and

Langley Research Center, NASA developed handouts that present facility-relevant climate change information in a user-friendly format to help decision makers at NASA centers understand what to expect in the future, so they can plan accordingly. To help nonscientists use the handout, it provides information on how to interpret local climate projections, identify specific potential impacts from climate change, and lays out the key adaptation considerations for local decision makers.

EPA and the Department of Transportation have also developed climate data translation resources for the road and bridge, and wastewater management system infrastructure categories, respectively, and support technical assistance efforts designed to help decision makers use these resources. EPA's Climate Ready Water Utilities initiative focuses on taking existing climate change science information and determining how it can be incorporated into planning for drinking water, wastewater, and stormwater infrastructure. The resources and tools developed under the Climate Ready Water Utilities initiative are designed for decision makers with different levels of adaptation experience, according to EPA officials. Decision makers with little experience can learn about adaptation options using EPA's *Adaptation Strategies Guide for Water Utilities*, while more advanced decision makers can use EPA planning tools to conduct a workshop or use EPA's Climate Resilience Evaluation and Awareness Tool, a risk assessment software tool that uses climate information from USGCRP's 2009 National Climate Assessment to enable utilities to evaluate a range of climate change scenarios from 2010 through 2090.[116] This tool allows decision makers to analyze how various adaptation strategies may help reduce climate risks, enabling them to prioritize the implementation of adaptation measures. In the future, according to EPA officials, the Climate Ready Water Utilities initiative will focus on developing tools for smaller utilities that have limited resources to engage technical experts for assistance. According to EPA officials, the agency has other projects under way focused on providing additional information and alternative approaches for communities. These projects include work on a decision-making framework to help decision makers select among different adaptation approaches, development of case studies to promote peer-to-peer learning on preparing for impacts, and development of a tool for users to evaluate options in a range of potential future water quality scenarios.

The Department of Transportation supports a range of technical assistance efforts focused on helping road and bridge infrastructure decision makers incorporate climate change information into planning processes. First, the department maintains the Transportation and Climate Change Clearinghouse, which provides access to existing literature on climate change adaptation and transportation issues, but less in the way of detailed site-specific information that decision makers need for infrastructure planning.[117] Second, the department, through its Federal Highway Administration, completed Phase 1 of the Gulf Coast Study in March 2008, which analyzed how changes in climate could affect transportation systems in the gulf coast region over the next 50 to 100 years.[118] A second phase of the Gulf Coast Study, scheduled to be completed in 2013 according to the Federal Highway Administration, is focusing on the Mobile, Alabama, region and will build on the information developed in Phase 1. The Phase 2 study inventoried critical infrastructure, assembled climate data and projections for the region, and will assess the vulnerability of the critical infrastructure across modes. The study will also develop transferrable tools and approaches that decision makers can use to determine which transportation systems most need to be protected and to identify and choose suitable adaptation options.[119]

The technical assistance and tools provided by EPA and the Department of Transportation hold promise as ways to help decision makers obtain the best available climate-related information for infrastructure planning. However, officials from EPA and the Department of Transportation said that they do not know the extent to which decision makers are using the tools they developed. EPA officials told us they were not sure about the extent to which utilities have used the agency's Climate Resilience Evaluation and Awareness Tool, and can only estimate the number of users by the times it has been downloaded and the number of participants in pilot programs and educational webinars. EPA officials told us that the agency plans to conduct additional outreach to decision makers. Likewise, according to Federal Highway Administration officials, the extent to which states and metropolitan planning organizations have used some of the agency's climate adaptation resources remains unclear. The officials said that the states and metropolitan planning organizations participating in pilot programs have used the agency's draft adaptation framework. In addition, federal officials track and collect feedback from the state and local agencies that have participated in the workshops and peer exchanges that the Federal Highway Administration has sponsored, according to agency officials. Importantly, a 2010 NRC report on informing decisions in a changing climate found it difficult to identify good reviews and clear unbiased discussions of the full range of decision support tools, their appropriate uses and limitations, and concluded that there could be a stronger role for the federal government to provide guidance on tools to support climate decisions, perhaps through a climate tools database, network, and best practice examples.

Accessing Local Assistance

At the locations we visited, having access to local assistance was a key variable that enabled decision makers to incorporate climate change into project level infrastructure planning. The entities coordinating federal adaptation efforts are beginning to reflect in strategic planning the need to develop and provide access to local expertise capable of bridging the gap between decision makers and scientists. For example, USGCRP's April 2012 strategic plan recognizes the need to improve the federal government's ability to translate climate information into what is needed by decision makers, and adaptation task force reports state that the federal government should enhance its capacity to translate information between scientists and decision makers.[120] The National Climate Assessment also provides an opportunity to engage with stakeholders and partners and is being structured to provide a continuing mechanism for engaging communities and networks of stakeholders at the local, state, tribal, and regional levels.

In addition to these initiatives, several federally sponsored "boundary organizations" have already attempted to fill the climate data translation gap between decision makers and scientists, albeit in a limited manner. These specialized organizations have, in some cases, proved instrumental in enabling scientists and users of scientific information to work productively together by improving communication, translation, and mediation between the two communities, according to a 2009 NRC report on informing decisions in a changing climate.[121] NOAA's Regional Integrated Science and Assessments (RISA) centers are examples of climate change boundary organizations, and they support research teams based in academic institutions that conduct interdisciplinary and regionally relevant research to inform resource management, planning, and public policy. RISA teams help build the nation's capacity to adapt to climate variability and change by providing information to local decision makers. For example, Seattle Public Utilities and King County recognized NOAA's local RISA program—the University of Washington Climate Impacts Group—as instrumental in helping to elevate the issue of climate change in the central Puget Sound region and Washington State.[122] As noted by CEQ, there are other examples of science-to-user continuums from which to learn, including U.S. Department of Agriculture Cooperative

Extension and NOAA Sea Grant Extension, which provide extension agents of all specializations with training in understanding and communicating climate change information to support adaptation.

However, according to several studies, future federal efforts could better focus on local climate change assistance. As stated in a 2011 NOAA RISA workshop report, effectively managing the interface between scientists and decision makers may require establishing a new field of science or career path either within the academic community or the federal government. Importantly, a 2012 NRC report on climate models notes that climate data translation

> needs to be done by qualified people to ensure that users receive the most accurate and appropriate information. The people currently doing this work come from a diversity of backgrounds such as weather modeling, engineering, statistics and environmental science. Currently, no standards exist for helping potential employers assess whether such people have the necessary skills in the appropriate use of climate model information to ensure that they can provide the most accurate and appropriate information to end users. This suggests an unmet need for training and accreditation programs in this area.

Accordingly, in the report, NRC recommended the development of a national education and accreditation program for "climate model interpreters" who can take technical findings and output from climate models, including quantified uncertainties, and use them in a diverse range of private- and public-sector applications. It is not clear what role the federal government could or should play in the development of such a program.

Whatever the federal role in the future of climate data translation, research and experience show that such activities are more effective when well-established organizations build trust among information users over time, and that, in many instances, formal institutionalization will be critical to longevity, recognition, and success, according to NRC.[123] The Interagency Climate Change Adaptation Task Force recognizes this need and stated, in its 2010 progress report, that to effectively integrate and implement adaptation responses, the federal government should recruit, develop, and retain technically capable staff that have the proper expertise to understand decision maker needs, and to communicate effectively the range of possible climate change impacts. USGCRP is also aware of this issue, noting in its April 2012 strategic plan, that USGCRP agencies will use their relationships with academia to promote the interdisciplinary education at undergraduate and graduate levels needed for a professional and technical workforce in areas related to climate change. These federal goals were developed too recently to evaluate, but it is unclear how developing a highly qualified workforce of climate interpreters without a corresponding institutional home would help infrastructure decision makers understand who they can contact for assistance.

Considering Climate Change in Existing Planning Processes

Notwithstanding the limited federal role in planning for transportation and wastewater infrastructure, several emerging federal adaptation efforts could help local infrastructure decision makers consider climate change in existing processes, according to studies, local site visit decision makers, and other stakeholders. These efforts relate to (1) design standards specifying how to consider climate change in infrastructure projects; (2) guidance specifying how certain types of federal infrastructure investments should account for climate change when meeting the requirements of the National Environmental Policy Act of 1969 (NEPA):[124] and (3) agency adaptation plans describing, among other things, how climate change will be considered in federal planning processes that influence local actions.

Design Standards

Professional associations like AASHTO—not federal agencies—generally develop the design standards that specify how weather and climate- related data are to be considered in project-level design and planning processes for roads and bridges, wastewater management systems, and NASA centers.[125] OMB Circular A-119 directs agencies to use these voluntary consensus standards in lieu of government-unique standards except where inconsistent with law or otherwise impractical.[126] According to Federal Highway Administration officials, for highway infrastructure these design standards are in turn modified and adopted by state governments and then approved by the federal government agency, in this case the Federal Highway Administration, before they can be applied to federally funded projects. Thus, federal agencies rely on professional associations to provide initial input to determine how and when climate- related data are included within design standards that specify how infrastructure is to be built.

Decision makers from the sites we visited, other infrastructure stakeholders, and relevant studies emphasized the importance of better employing design standards as a tool for incorporating climate change in infrastructure planning. For example, experts from the University of

Washington who work with the King County Wastewater Treatment Division stated that it would be helpful to have (1) protocols for developing and maintaining design standards that incorporate climate change projections and (2) established methods for using this information in actual design processes via well-documented case studies; because, according to these experts, not having a formal process for incorporating climate change information in design standards effectively ensures that most of the design community cannot act without unacceptable professional risks. Similarly, officials from the American Society of Civil Engineers with whom we spoke acknowledged that incorporating climate science in design standards is critical for translating adaptation into engineering practice. Building on this point, a recent report on adaptation policy noted that updating design standards can also spur innovation in materials science, engineering, and construction.[127]

Professional associations are beginning to take interest in climate change adaptation. For example, AASHTO maintains a web-based Transportation and Climate Change Resource Center with a climate adaptation page and a list of educational webinars on topics such as adapting infrastructure to extreme events.[128] Also, in 2011, the American Society of Civil Engineers developed a Committee on Adaptation to Climate Change to, in part, translate climate science into engineering practice. In addition, some private infrastructure development and construction companies are beginning to develop methods to compare the costs and benefits of engineering alternatives considering different climate futures. These efforts are just under way, with as yet undetermined outcomes, but, according to a TRB-commissioned study, updating standards is a long process, involving many government and nongovernmental standard-setting organizations.

As a result, there have been calls for a more active federal role in encouraging professional associations to consider climate change in design standards.[129] In 2010, NRC identified as a national priority the revision of engineering standards to reflect current and anticipated future climate changes, and it recommended that their use be required as a condition for federal investments in infrastructure.[130] While not going as far as the NRC recommendation, recent transportation legislation recognized the significance of design standards. Section 33009 in the Senate version of the Moving Ahead for Progress in the 21st Century Act would have required the Secretary of Transportation, in consultation with others, to issue guidance and establish design standards for transportation infrastructure to help states and other entities plan for natural disasters and a greater frequency of extreme weather events in the process of planning, siting, designing, and developing transportation infrastructure by assessing vulnerabilities to a changing climate and the costs and benefits of adaptation measures.[131] Section 33009 was not, however, in the version of the bill the conference committee agreed to, which ultimately passed both Houses of Congress and was signed into law on July 6, 2012.

NEPA Guidance

Certain types of federal infrastructure investments need to meet the requirements of NEPA, which requires federal agencies to evaluate the environmental impacts of their proposed actions and reasonable alternatives to those actions. Usually federal agencies evaluate the likely environmental effects of major federal actions using an environmental assessment, or, if the action likely would significantly affect the environment, a more detailed environmental impact statement.[132]

On February 18, 2010, CEQ—the entity within the Executive Office of the President that oversees implementation of NEPA—issued draft guidance on how federal agencies can consider the effects of climate change in the NEPA process.[133] As CEQ noted in this guidance, the environmental analysis and documents produced in the NEPA process could consider the relationship of climate change effects to a proposed action, such as an infrastructure project that was a major federal action, or alternatives, including proposal design and adaptation measures.

CEQ's draft NEPA guidance states that climate change effects should be considered in the analysis of projects that are designed for long-term utility and located in areas that are considered vulnerable to specific effects of climate change (e.g., increasing sea level or ecological change) within the project's time frame. For example, a proposal for long-term development of transportation infrastructure on a coastal barrier island will likely need to consider whether environmental effects or design parameters may be changed by the projected increase in the rate of sea level rise. Given the length of time involved in present sea level projections, such considerations typically would not be relevant to an action with only short-term considerations. The guidance further states that this is not intended as a new component of NEPA analysis but rather as a potentially important factor to be considered within the existing NEPA framework. The draft guidance also noted that, after consideration of public comment, CEQ intended to expeditiously issue the guidance in final form. CEQ received public comments on the draft guidance following its release on February 18, 2010. CEQ has not finalized the guidance or issued regulations addressing how, if at all, federal agencies are to consider the effects of climate change in the NEPA process. When asked for an estimate on when the final guidance would be available, CEQ, in December 2012, stated that "we are continuing to assess the best approach moving forward as we work on developing the guidance," but did not indicate when the guidance would be finalized. [134] Without finalized guidance from CEQ, it is unclear how, if at all, agencies are to consistently consider climate change in the NEPA process, creating the potential for inconsistent consideration of the effects of climate change in the NEPA process across the federal government.

Agency Adaptation Plans

As directed by CEQ instructions and guidance implementing Executive Order 13514, agency adaptation plans for fiscal year 2013 were submitted to CEQ in June 2012 as part of executive branch agencies' annual Strategic Sustainability Performance Plans. According to CEQ, the adaptation plans are to outline the agency's policy framework, analysis of climate change risks and opportunities, process for agency adaptation planning and evaluation, programmatic activities, and actions taken to better understand and address the vulnerabilities posed by a changing climate. Agencies are to consider how they will include climate change within their existing programs and planning processes, some of which can influence state and local actions on infrastructure investment. For example, on September 24 2012, the Federal Highway Administration's Associate Administrators for Infrastructure; Planning, Environment, and Realty; and Federal Lands Highway issued a memorandum to Federal Highway Administration staff clarifying the eligibility of adaptation activities for federal highway funding. The memo notes that Federal Highway Administration offices may allow state and local agencies to use highway funds to consider the potential impacts of climate change and extreme weather events and apply adaptation strategies, both at the project and systems levels. The extent to which agency adaptation plans will address policy specifics such as the Federal Highway Administration guidance is unclear because draft plans were released on February 7, 2013, and are undergoing public review and comment.[135]

CONCLUSION

Physical infrastructure such as roads, bridges, wastewater management systems, and NASA centers are typically expensive and long-term federally funded investments. Many are projected to be impacted by changes in the climate that, according to best available science, are inevitable in coming decades. As the nation makes these investments, it faces the choice of paying more now to account for the risk of climate change, or potentially paying a much larger premium later to repair, modify, or replace infrastructure ill-suited for future conditions. The choice raises a basic risk management question that an increasing number of state and local decision makers are beginning to address, particularly in the aftermath of Superstorm Sandy.

Planning for transportation and wastewater infrastructure in this country remains largely within the domain of state and local governments, but emerging federal efforts are under way to facilitate and enable more informed decisions about adaptation. Moreover, entities coordinating federal adaptation efforts are beginning to reflect in strategic planning the need to develop and provide access to local assistance capable of bridging the gap between decision makers and scientists. Studies, local decision makers from site visits, and stakeholders suggest ways federal adaptation efforts could better serve the needs of local infrastructure decision makers. Specifically:

- Federal agencies and academic institutions collect a vast array of climate-related data, but local infrastructure decision makers face difficulty identifying, accessing, and using them, because as noted by a 2010 NRC study, this information exists in an uncoordinated confederation of networks and institutions. Of particular note, federal efforts to provide access to site-specific, climate-related information are an emerging priority, but it remains unclear how these efforts will address the challenge of identifying the best available information to use in infrastructure planning. According to the 2010 NRC report, the end result of this information not being easily accessible is that people may make decisions—or choose not to act—without it.
- At the locations we visited, access to local assistance was a key variable that enabled decision makers to translate available climate-related data into information useful for project level planning, but it is unclear how emerging federal efforts will help decision makers in other locations obtain similar assistance. Without clear sources of local assistance, infrastructure decision makers—who may not be familiar with climate science and who have many other responsibilities of immediate importance—will be left to sort it out themselves, and will face difficulty justifying investment in adaptation measures, the benefits of which may not be realized for several decades into the future.

- Notwithstanding the limited role federal agencies play in most project- level planning, certain types of federal infrastructure investments need to meet the requirements of NEPA. On February 18, 2010, CEQ issued draft guidance on how federal agencies can consider the effects of climate change in the NEPA process. However, CEQ has not finalized the guidance or issued regulations addressing how, if at all, federal agencies are to consider the effects of climate change in the NEPA process, and it also has not indicated when or if the guidance would be finalized. Without finalized guidance from CEQ, it is unclear how, if at all, agencies are to consistently consider climate change in the NEPA process.
- Professional associations generally develop and maintain design standards critical for translating adaptation into infrastructure engineering practice, not relevant federal agencies such as EPA (which has the lead for federally funded wastewater systems) or the U.S. Department of Transportation (which has the lead for federally funded roads and bridges). OMB Circular A-119 directs federal agencies to use voluntary consensus standards in lieu of government-unique standards except where inconsistent with law or otherwise impractical. Professional associations have started to investigate how to incorporate climate-related data into design standards, with as yet undetermined outcomes. Not having a formal process for incorporating climate change information in design standards effectively ensures that most of the infrastructure design community cannot act without unacceptable professional risks, according to certain local decision makers and stakeholders. As a result, there have been calls for a more active federal role in encouraging professional associations to consider climate change in design standards.

RECOMMENDATIONS FOR EXECUTIVE ACTION

To improve the resilience of the nation's infrastructure to climate change, we are making the following four recommendations:

- that the Executive Director of the United States Global Change Research Program or other federal entity designated by the Executive Office of the President work with relevant agencies to

  - identify for decision makers the "best available" climate-related information for infrastructure planning and update this information over time and
  - clarify sources of local assistance for incorporating climate-related information and analysis into infrastructure planning, and communicate how such assistance will be provided over time;
- that the Chairman of the Council on Environmental Quality finalize guidance on how federal agencies can consider the effects of climate change in their evaluations of proposed federal actions under the National Environmental Policy Act; and
- that the Secretary of the U.S. Department of Transportation and the Administrator of the Environmental Protection Agency work with relevant professional associations to incorporate climate change information into design standards.

AGENCY COMMENTS

We provided a draft of this report for review and comment to the Secretary of Transportation, the Administrator of EPA, the Chair of CEQ, the Director of OSTP, and the Executive Director of USGCRP. They did not provide official written comments but instead provided technical comments, which we incorporated, as appropriate.

David C. Trimble
Director, Natural Resources and Environment

APPENDIX I: OBJECTIVES, SCOPE, AND METHODOLOGY

This report (1) describes what is known about the impacts of climate change on the nation's infrastructure, specifically roads and bridges, wastewater management systems, and National Aeronautics and Space Administration (NASA) centers; (2) analyzes the extent to which potential climate change impacts are incorporated into infrastructure planning; (3) identifies the factors that enabled certain decision makers to integrate climate change impacts into infrastructure planning; and (4) analyzes federal efforts to address the adaptation needs of local infrastructure decision makers and describes potential opportunities for improvement identified by studies, local decision makers who integrated climate change into infrastructure planning, and other stakeholders. We selected the road and bridge and wastewater management system infrastructure categories because they account for significant federal funding and are the focus of specific federal adaptation initiatives.[136] We selected NASA centers because these facilities are large, they manage mission critical assets that are difficult, if not impossible to move or replace and, importantly, NASA has an emerging partnership effort focused on considering climate change information within the planning for its centers. NASA centers are also instructive examples because they incorporate roads, bridges, wastewater systems, and other infrastructure in one place as a system to support a mission.

Before describing in detail the methods we used, it is important to recognize a few limits of our approach and report. First, it focuses on planning for new projects or significant rebuilds, and does not focus on operations and maintenance or wide-scale efforts to assess the vulnerability of the existing stock of infrastructure. Second, this report focuses on planning for specific projects, not long-range planning or strategic prioritization processes. Third, this report describes how decision makers incorporated climate change adaptation into infrastructure planning and implementation, but it does not generally assess the effectiveness of the adaptive actions themselves. The need for further research in this area is widely acknowledged but is not the focus of this report.

To explain the potential consequences of climate change on the Nation's infrastructure, we reviewed assessments from the National Research Council, the United States Global Change Research Program, and relevant federal agencies. We identified these assessments using government and National Academies websites and prior GAO reports on climate change. We then evaluated whether the assessments fit within the scope of work and contributed to the objectives of this report. For relevant assessments, we used in-house scientific expertise to analyze the soundness of the methodological approaches they utilized, and we determined them to be sufficiently sound for our purposes. Relevant assessments are cited throughout this document.

To identify the extent to which climate change impacts are incorporated into infrastructure planning, we (1) reviewed laws, regulations, and planning guidance; (2) analyzed relevant reports on climate change adaptation; and (3) interviewed knowledgeable infrastructure stakeholders and officials from professional associations, federal agencies, and other organizations. To identify relevant reports on climate change adaptation, we conducted a literature search and review with the assistance of a technical librarian. We searched various databases, such as ProQuest, and focused on peer reviewed journals, government reports, trade and industry articles, and publications from associations, nonprofits, and think tanks from 2005 to present. We also searched for reports from the Congressional Research Service, the Congressional Budget Office, and agency inspectors general. To supplement this review we analyzed Internet-based adaptation report databases such as the Climate Adaptation Knowledge Exchange.[137] Relevant reports are cited in footnotes throughout this report. To identify knowledgeable stakeholders, we reviewed our prior climate change work and relevant reports to identify individuals with specific knowledge of climate change adaptation and infrastructure. We interviewed professional association stakeholders from the American Association of State Highway and Transportation Officials, American Society of Civil Engineers, National Association of Clean Water Agencies, and the Water Utility Climate Alliance; federal agency officials from the Environmental Protection Agency and the Federal Highway Administration; and other stakeholders familiar with infrastructure adaptation, including the Georgetown Climate Center and the Center for Climate and Energy Solutions. We also coordinated with the Congressional Budget Office and the Congressional Research Service.

To examine how climate change has been considered in infrastructure planning, we visited seven locations where decision makers had done so—three locations focused on roads and bridges (Washington State Route 522; Interstate-10 Twin Span Bridge near New Orleans, Louisiana; and Louisiana State Highway 1) , two locations focused on wastewater management systems category (King County Wastewater Treatment Division in Washington and the Milwaukee Metropolitan Sewerage District in Wisconsin), and two NASA centers (Langley Research Center in Hampton, Virginia, and Johnson Space Center in Houston, Texas). To select the transportation and wastewater sites, we reviewed studies; interviewed federal, state, and local agency officials; and analyzed Internet-based adaptation case study databases maintained by academic institutions such as the Georgetown Climate Center to identify examples where climate change was considered in infrastructure planning.[138] From this review, we found a universe of about 20 total potential transportation and wastewater management system examples. Based on follow-up interviews and additional research, we narrowed the potential list for each category based on whether the candidates had considered climate change during both the project planning and implementation phases. We selected three projects focused on roads and bridges and two locations focused on wastewater management systems in an attempt to illustrate different potential climate impacts in different regions of the United States (the Pacific Northwest, Great Lakes, Gulf Coast, and Mid-Atlantic), but we were somewhat limited by the small set of potential site visits. NASA scheduled climate change adaptation workshops at two of its centers (Langley Research Center and Johnson Space Center) during the time frame of our work. We attended the workshops and collected information from a variety of federal and local stakeholders, including government officials and academic institutions. The sites we selected are not representative of all infrastructure adaptation efforts taking place; however, they include a variety of responses to climate change effects across different infrastructure categories. Findings from these site visits cannot be generalized to those we did not include in our nonprobability sample.

We gathered information during and after the site visits through observation of adaptation efforts, interviews with officials and stakeholders, and a review of documents provided by these officials. As part of the site visits, we interviewed academic institutions that provided climate-related information to decision makers, including the Wisconsin Initiative on Climate Change Impacts, a collaboration between the University of Wisconsin–Madison's Nelson Institute for Environmental Studies and the Wisconsin Department of Natural Resources; the Climate Impacts Group, an interdisciplinary research group at the University of Washington; and the Southern Climate Impacts Planning Program, a collaborative research program of the University of Oklahoma and Louisiana State University. We also followed up with officials after our visits to gather additional information.

To analyze federal efforts to address the adaptation needs of state and local infrastructure decision makers and to describe opportunities for improvement, we (1) interviewed federal officials from the Council on Environmental Quality, Department of Transportation's Federal Highway Administration, Environmental Protection Agency, and United States Global Change Research Program and (2) reviewed available studies on federal adaptation efforts. To monitor federal adaptation-related activities, we accessed materials stored in www.fedcenter.gov, the federal government's home for comprehensive environmental stewardship and compliance assistance information. We also attended the Adaptation Futures International Conference on Climate Adaptation in May 2012 cohosted by the University of Arizona in Tucson, Arizona, and by the United Nation Environment Programme's Programme of Research on Climate Change Vulnerability,

Impacts and Adaptation, to learn about climate change adaptation research and approaches from around the world.

We conducted this performance audit from October 2011 to April 2013 in accordance with generally accepted government auditing standards. Those standards require that we plan and perform the audit to obtain sufficient, appropriate evidence to provide a reasonable basis for our findings and conclusions based on our audit objectives. We believe that the evidence obtained provides a reasonable basis for our findings and conclusions based on our audit objectives.

End Notes

[1] Additional information on billion dollar weather disasters is available at NOAA's National Climatic Data Center.

[2] USGCRP coordinates and integrates the activities of 13 federal agencies that conduct research on changes in the global environment and their implications for society. USGCRP began as a presidential initiative in 1989 and was codified in the Global Change Research Act of 1990 (Pub. L. No. 101-606, § 103 (1990)). USGCRP-participating agencies are the Departments of Agriculture, Commerce, Defense, Energy, Interior, Health and Human Services, State, and Transportation; the U.S. Agency for International Development; the Environmental Protection Agency; the National Aeronautics and Space Administration; the National Science Foundation; and the Smithsonian Institution.

[3] Thomas R. Karl, Jerry M. Melillo, and Thomas C. Peterson, eds. *Global Climate Change Impacts in the United States* (New York, NY: Cambridge University Press, 2009). This document, referred to as the 2009 National Climate Assessment, is in the process of being updated, as the Global Change Research Act of 1990 requires that a scientific assessment be provided to the President and Congress not less frequently than every 4 years.

[4] NRC is the operating arm of the National Academy of Sciences and National Academy of Engineering. Through its independent, expert reports; workshops; and other scientific activities, NRC's mission is to improve government decision making and public policy, increase public understanding, and promote the acquisition and dissemination of knowledge in matters involving science, engineering, technology, and health.

[5] There are multiple definitions of climate change adaptation. For example, in USGCRP's 2009 assessment, adaptation refers to changes made to better respond to present or future climatic and other environmental conditions, thereby reducing harm or taking advantage of opportunity. Various federal officials commented that this definition better emphasizes that adaptation is meant not just to be an adjustment in response to climate change, but a change in the way of thinking so as to reduce harm and take advantage of opportunity.

[6] GAO, *Climate Change Adaptation: Strategic Federal Planning Could Help Government Officials Make More Informed Decisions,* GAO-10-113 (Washington, D.C.: Oct. 7, 2009).

[7] Department of Defense, *Quadrennial Defense Review Report* (Washington, D.C.: February 2010.

[8] Department of Homeland Security, *Quadrennial Homeland Security Review Report: A Strategic Framework for a Secure Homeland* (Washington, D.C.: February 2010).

[9] GAO, *High-Risk Series: An Update,* GAO-13-283, February 2013. Every 2 years at the start of a new Congress, GAO calls attention to agencies and program areas that are high risk due to

their vulnerabilities to fraud, waste, abuse, and mismanagement, or are most in need of transformation.

[10] According to the NRC's 2007 *Analysis of Global Change Assessments: Lessons Learned*, scientific assessments are evaluation and consensus building processes for establishing an integrated view of recent scientific breakthroughs and providing policy- relevant information to decision makers. An assessment can establish the importance of an issue, provide an authoritative resolution of policy-relevant scientific questions, demonstrate the benefits of policy options, identify new research directions, and provide technical solutions. The assessment process in itself is a key interface between science and policy and a crucial mechanism by which science informs policy making. For assessments to be effective and credible, the process has to be open and must provide accurate, useful, and scientifically tested information.

[11] This report focuses on distinct types of infrastructure. We have additional climate change adaptation work under way focused on different types of infrastructure. For example, we expect to complete work on energy and water infrastructure adaptation and how climate change is considered in federal natural resource planning.

[12] Because this was a nonprobability sample, the information collected during these site visits cannot be generalized to all U.S. infrastructure planning but provides illustrative information for locations in which such planning has been undertaken.

[13] NRC, America's Climate Choices: Panel on Adapting to the Impacts of Climate Change, *Adapting to the Impacts of Climate Change* (Washington, D.C.: 2010).

[14] GAO, *Physical Infrastructure: Challenges and Investment Options for the Nation's Infrastructure,* GAO-08-763T (Washington, D.C.: May 8, 2008).

[15] Congressional Budget Office, *Public Spending on Transportation and Water Infrastructure*, Pub. No. 4088 (Washington, D.C.: November 2010).

[16] For more information on EPA's Clean Water State Revolving, Visit http://water.epa.gov/grants_funding/cwsrf/cwsrf_index.cfm.

[17] For more information, see the General Services Administration's Federal Real Property Profile inventory system, Visit http://www.gsa.gov/portal/content/104199.

[18] NASA Office of the Inspector General, *NASA'S Infrastructure and Facilities: An Assessment of the Agency's Real Property Master Planning*. REPORT NO. IG-12-008 (Washington, D.C.: Dec. 19, 2011).

[19] The American Society of Civil Engineers represents more than 140,000 members of the civil engineering profession worldwide.

[20] NASA Office of the Inspector General, *NASA'S Infrastructure and Facilities: An Assessment of the Agency's Real Property Master Planning*. REPORT NO. IG-12-008 (Washington, D.C.: Dec. 19, 2011).

[21] GAO, *Wastewater Infrastructure Financing: Stakeholder Views on a National Infrastructure Bank and Public-Private Partnerships*, GAO-10-728 (Washington, D.C.: June 30, 2010).

[22] For more information about the American Society of Civil Engineers' Report Card for America's Infrastructure, Visit http://www.infrastructurereportcard.org/.

[23] GAO, *Physical Infrastructure: Challenges and Investment Options for the Nation's Infrastructure*, GAO-08-763T (Washington, D.C.: May 8, 2008). See also GAO, *Transportation: Key Issues and Management Challenges*, GAO-12-581T (Washington, D.C.: Mar. 29, 2012).

[24] Daniel F. Morris, Molly K. Macauley, Raymond J. Kopp, and Richard D. Morgenstern, Resources for the Future, *Reforming Institutions and Managing Extremes: U.S. Policy Approaches for Adapting to a Changing Climate* (Washington D.C.: 2011).

[25] The National Academies, Committee on Increasing National Resilience to Hazards and Disasters; Committee on Science, Engineering, and Public Policy; *Disaster Resilience: A National Imperative* (Washington, D.C., 2012).

[26] NRC, America's Climate Choices: Panel on Adapting to the Impacts of Climate Change, *Adapting to the Impacts of Climate Change* (Washington, D.C.: 2010).

[27] GAO

[28] Center for Climate and Energy Solutions, *Extreme Weather & Climate Change: Understanding the Link and Managing the Risk* (December 2011).

[29] ISO is a worldwide federation of national standards bodies. Please also note that under the *Standards for Internal Control in the Federal Government* (GAO) the risk assessment standard states that federal agencies are to provide for an assessment of the risks they face from both external and internal sources and that because governmental, economic, industry, regulatory, and operating conditions continually change, mechanisms should be provided to identify and deal with any special risks prompted by such changes.

[30] ISO 31000:2009 *Risk management – Principles and guidelines* is available at http://www.iso.org/iso/catalogue_detail?csnumber=43170.

[31] Office of Management and Budget, *Instructions for the Planning, Budgeting, Acquisition and Management of Non-IT Capital Assets,* fiscal year 2013.

[32] In technical comments, CEQ, OSTP, and USGCRP noted that to the extent federal funds are involved, federal agencies may have authority to either incentivize or require states and local governments to take specific adaptation actions or adopt risk management strategies but that more opportunities to incentivize adaptation planning may exist and, in some cases, additional statutory authority is necessary to require action.

[33] GAO, *Highway Infrastructure: Federal-State Partnership Produces Benefits and Poses Oversight Risks*, GAO-12-474 (Washington, D.C.: Apr. 26, 2012).

[34] GAO, *Federal-Aid Highways: FHWA Needs a Comprehensive Approach to Improving Project Oversight*, GAO-05-173 (Washington, D.C.: Jan. 31, 2005). Specifically, statewide and metropolitan planning organizations' transportation plans and improvement programs provide for development and management of transportation systems and facilities.

[35] Pub. L. No. 112-141 (2012). In technical comments, the Department of Transportation noted that it was considering how to ensure that rebuilding efforts reflect sea level rise projections and expected flood levels in implementing the Public Transportation Emergency Relief Program authorized by the Moving Ahead for Progress in the 21st Century Act. In addition, the Disaster Relief Appropriations Act, 2013 appropriated funds for highway and transportation projects related to reducing risk of damage from future disasters in areas impacted by Hurricane Sandy.

[36] Pub. L. No. 111-5. 123 Stat. 115. 169 (2009). A similar provision was included in EPA's fiscal year 2010 and 2011 appropriations, and in the agency's fiscal year 2012 appropriation, although that law required not less than 10 percent of the appropriation be used for such eligible projects.

[37] In addition, on January 29, 2013, $500 million was appropriated for the Clean Water State Revolving Fund program for eligible projects in EPA Region 2—which includes New Jersey and New York—whose purpose is to reduce flood damage risk and vulnerability or to enhance resiliency to rapid hydrologic change or a natural disaster at a drinking water or wastewater facility.

[38] Milwaukee Metropolitan Sewerage District officials we spoke with noted the potential of EPA's green infrastructure activities to encourage adaptive action, particularly a rulemaking to strengthen the agency's stormwater program. According to agency officials, EPA intends to propose a rule to strengthen the stormwater program by June 10, 2013, and complete a final rule by December 10, 2014.

[39] See, for example, NRC, Panel on Strategies and Methods for Climate-Related Decision Support, Committee on the Human Dimensions of Global Change, *Informing Decisions in a Changing Climate* (Washington, D.C.: 2009).

[40] Thomas R. Karl, Jerry M. Melillo, and Thomas C. Peterson, eds., *Global Climate Change Impacts in the United States* (New York, NY: Cambridge University Press, 2009).

[41] NRC, Committee on Climate Change and U.S. Transportation, TRB and Division on Earth and Life Studies, *Potential Impacts of Climate Change on U.S. Transportation* (Washington, D.C.: 2008).

[42] In technical comments, EPA noted that flooding, erosion, and scouring are not just problems in coastal areas and can also be caused by more intense precipitation combined with impervious cover, resulting in higher velocity stream flows.

[43] U.S. Climate Change Science *Program, Impacts of Climate Change and Variability on Transportation Systems and Infrastructure: Gulf Coast Study, Phase I.* A Report by the U.S. Climate Change Science Program and the Subcommittee on Global Change Research (Savonis, M. J., V.R. Burkett, and J.R. Potter [eds.]). Department of Transportation (Washington, D.C.: 2008).

[44] Department of Homeland Security, National Infrastructure Simulation and Analysis Center, Risk Development and Modeling Branch, Homeland Infrastructure Threat and Risk Analysis Center, Office of Infrastructure Protection, In Collaboration with The National Incident Management Systems and Advanced Technologies Institute at The University of Louisiana at Lafayette, *Louisiana Highway 1/Port Fourchon Study* (July 15, 2011).

[45] EPA, *National Water Program 2012 Strategy: Response to Climate Change* (December 2012); National Association of Clean Water Agencies and Association of Metropolitan Water Agencies, *Confronting Climate Change: An Early Analysis of Water and Wastewater Adaptation Costs* (October 2009); Thomas R. Karl, Jerry M. Melillo, and Thomas C. Peterson, eds., *Global Climate Change Impacts in the United States* (New York, NY: Cambridge University Press, 2009).

[46] Wastewater Committee of the Great Lakes—Upper Mississippi River Board of State and Provincial Public Health and Environmental Managers. *Recommended Standards for Wastewater Facilities—2004 Edition* (Albany, NY: 2004). According to EPA, many states follow this "10-state standard."

[47] As EPA states on its combined sewer overflow web page, combined sewer systems are sewers that are designed to collect rainwater runoff, domestic sewage, and industrial wastewater in the same pipe. Most of the time, combined sewer systems transport all of their wastewater to a sewage treatment plant, where it is treated and then discharged to a water body.

[48] The Milwaukee Metropolitan Sewerage District provides wastewater and flood management services to about 1.1 million customers in the Milwaukee region in Wisconsin.

[49] In 2008, 2009, and 2010, the Milwaukee region experienced storms with rainfall totals that had less than a 1 percent probability of occurring at that location that year—in other words, storms that exceeded the 100-year storm. For example, the July 2010 storm exceeded the 700-year storm, which means it had about a 0.14 percent probability of occurring at that location.

[50] Unless otherwise noted, the content of the "Johnson Space Center" section of this report is based on information presented by NASA officials on March 6-8, 2012, as part of the Johnson Space Center resilience and adaptation to climate risks workshop.

[51] Unless otherwise noted, the content of the "Langley Research Center" section of this report is based on information presented by NASA officials on September 27-29, 2011, as part of the Langley Research Center resilience and adaptation to climate risks workshop.

[52] According to its website, the Strategic Environmental Research and Development Program is DOD's environmental science and technology program, executed in partnership with the Department of Energy and EPA. The program invests in basic and applied research and advanced development. For more information on the Risk Quantification for Sustaining Coastal Military Assets and Mission Capabilities study, please visit http://www.serdp.org/Program-Areas/Resource-Conservation-and-Climate-Change/Climate-Change/Vulnerability-and-Impact-Assessment/RC-1701. For a related study, see also NRC, Naval Studies Board, Committee on *National Security Implications of Climate Change for U.S. Naval Forces*; National Security Implications of Climate Change for U.S. Naval Forces (Washington, D.C.: 2011).

[53] We also reported on this challenge in GAO

[54] NRC, Committee on Climate Change and U.S. Transportation, TRB and Division on Earth and Life Studies, *Potential Impacts of Climate Change on U.S. Transportation* (Washington,

D.C.: 2008). National Drinking Water Advisory Council, *Climate Ready Water Utilities Final Report* (Dec. 9, 2010).

[55] EPA is taking steps in preparation for proposing a rule to establish a program to reduce stormwater discharges from newly developed and redeveloped sites and make other regulatory improvements to strengthen its stormwater program. According to EPA officials, the agency will propose the rule by June 10, 2013. National Drinking Water Advisory Council, *Climate Ready Water Utilities Final Report* (Dec. 9, 2010).

[56] According to technical comments from CEQ, OSTP, and USGCRP, several of the examples listed in this section (meeting permit requirements, upgrading wastewater treatment plants, and preparing to implement proposed stormwater rules) are driven by federal agencies and federal actions, leading to a way federal agencies could incentivize and encourage more of a focus on this issue.

[57] NRC, Committee on Climate Change and U.S. Transportation, TRB and Division on Earth and Life Studies, *Potential Impacts of Climate Change on U.S. Transportation* (Washington, D.C.: 2008).

[58] In technical comments, CEQ, OSTP, and USGCRP noted that this paragraph is a great illustration of the misperception that all climate change impacts will occur gradually. Evidence is mounting rapidly that extreme events associated with climate change are already occurring and that the costs are increasing.

[59] NASA officials commented that the statement "decision makers face challenges obtaining the best available climate-related information for infrastructure planning" does not apply to decision makers at its facilities. NASA's efforts to share climate-related information with decision makers at its centers are described in more detail later in this report.

[60] NRC, Committee on *a National Strategy for Advancing Climate Modeling*, Board on Atmospheric Studies and Climate, Division on Earth and Life Sciences, A National Strategy for Advancing Climate Modeling (Washington, D.C.: 2012).

[61] NRC, America's Climate Choices: Panel on Informing Effective Decisions and Actions Related to Climate Change, *Informing an Effective Response to Climate Change* (Washington, D.C.: 2010).

[62] According to CEQ, OSTP, and USGCRP technical comments, the issue often is figuring out which information is relevant to a particular decision context, not so much the quality of the information itself.

[63] In technical comments, the Department of Transportation noted that climate impacts are not equally relevant to all projects and not all climate impacts are sufficiently predictable at a local scale to meaningfully analyze. Sea level rise, for instance, is relatively straightforward, while predicting fresh-water flooding in a particular watershed due to changes in precipitation may be very challenging.

[64] NRC, America's Climate Choices: Panel on Informing Effective Decisions and Actions Related to Climate Change, *Informing an Effective Response to Climate Change* (Washington, D.C.: 2010).

[65] This statement is not intended to suggest that regional coordinated adaptation actions are not necessary or useful, but instead to emphasize the importance of locally relevant information for decision making.

[66] NRC, America's Climate Choices: Panel on Informing Effective Decisions and Actions Related to Climate Change, *Informing an Effective Response to Climate Change* (Washington, D.C.: 2010).

[67] NRC, Committee on *a National Strategy for Advancing Climate Modeling*, Board on Atmospheric Studies and Climate, Division on Earth and Life Sciences, A National Strategy for Advancing Climate Modeling (Washington, D.C.: 2012).

[68] NOAA Climate Program Office, *RISA Workshop Report: Looking ahead at climate service, assessment, and adaptation* (Silver Spring, MD: 2011).

[69] Environmental and Energy Study institute and the Center for Clean Air Policy, *Climate Adaptation & Transportation: Identifying Information and Assistance Needs*, Prepared for

the NOAA Sector Applications and Research Program (May 2012). The report went on to describe the challenge more succinctly, "As Bill Murray and Dan Aykroyd put it: Who Ya Gonna Call?"

[70] NRC, Panel on Strategies and Methods for Climate-Related Decision Support, Committee on the Human Dimensions of Global Change, *Informing Decisions in a Changing Climate* (Washington, D.C.: 2009).

[71] In technical comments, CEQ, OSTP, and USGCRP emphasized that although downscaled information would be useful, lack of such information should not be a barrier for adaptation. Decision makers can still move forward without downscaled models or science. In addition, uncertainty is not a reason to do nothing because planners make decisions under all kinds of other uncertainties, including development patterns and demographics, according to the technical comments.

[72] CEQ, OSTP, and USGCRP commented that engineering processes are designed to handle a predictable amount of variability. Climate change projections pose complications to this existing process of accounting for variability, as climate change exacerbates the extremes of what has been accounted for historically.

[73] NRC, Committee on *a National Strategy for Advancing Climate Modeling*, Board on Atmospheric Studies and Climate, Division on Earth and Life Sciences, A National Strategy for Advancing Climate Modeling (Washington, D.C.: 2012).

[74] Design standards define how a product is to be designed and built by providing technical guidelines that promote the safety, reliability, productivity, and efficiency of infrastructure. Design standards for each infrastructure category are typically developed by relevant professional associations.

[75] These statistical tables are known as "Atlas 14." According to NOAA's documentation, NOAA Atlas 14 contains precipitation frequency estimates with associated confidence limits for the United States and is accompanied by additional information such as temporal distributions and seasonality. The Atlas is divided into volumes based on geographic sections of the country. The Atlas is intended as the official documentation of precipitation frequency estimates and associated information for the United States. It includes discussion of the development methodology and intermediate results.

[76] NRC, America's Climate Choices: Panel on *Adapting to the Impacts of Climate Change*, Adapting to the Impacts of Climate Change (Washington, D.C.: 2010).

[77] The project was fully funded by the Department of Transportation's Emergency Relief Program.

[78] Efforts to raise the northern section, which connects residents of the levee district with their work sites at Port Fourchon or offshore, are currently unfunded.

[79] In technical comments, CEQ, OSTP, and USGCRP noted that the Louisiana State Highway 1 example demonstrates in near-real time the benefits of specific adaptations in design. Having the old road adjacent to the new one and documenting the impacts to both is a great start in moving towards understanding the benefits of investments in adaptation as opposed to investing in disaster recovery, according to the comments.

[80] Mitigation banking is wetland restoration, creation, enhancement, or in certain circumstances, preservation undertaken for the purpose of compensating for unavoidable wetland losses.

[81] The Barton Pump Station overflow weir controls flows by setting the height at which incoming combined sewage can collect in a well when its flow exceeds the capacity of the pumping station. Once that height is exceeded, combined sewage overtops the weir, and excess flows are discharged to Puget Sound. Flap gates help prevent backflow into pumping stations.

[82] The Milwaukee Metropolitan Sewerage District purchases property as a part of its Greenseams® Program.

[83] According to technical comments from CEQ, OSTP, and USGCRP, a question now on the table is whether the experience of Superstorm Sandy and Hurricane Katrina is now allowing people across the country to experience these disasters indirectly and transfer knowledge to their own cases rather than having to personally experience disasters in order to be

motivated. It appears that interest in adaptation is increasing rapidly in the post-Sandy world.

[84] A 100-year storm means that the storm had less than a one percent probability of occurring at that location that year.

[85] NRC, America's Climate Choices: Panel on Adapting to the Impacts of Climate Change, *Adapting to the Impacts of Climate Change* (Washington, D.C.: 2010).

[86] NRC, Panel on Strategies and Methods for Climate-Related Decision Support, Committee on the Human Dimensions of Global Change, *Informing Decisions in a Changing Climate* (Washington, D.C.: 2009).

[87] According to technical comments from CEQ, OSTP, and USGCRP, policy documents are important because they institutionalize priorities and action.

[88] Uncertainty is an important consideration in discussions about how to address climate change. Take, for example, the recent decision by the United States Court of Appeals for the District of Columbia Circuit on EPA's rules regarding regulation of greenhouse gases under the Clean Air Act, including the finding that greenhouse gas emissions from motor vehicles is reasonably anticipated to endanger public health and welfare. The decision notes that the "existence of some uncertainty does not, without more, warrant invalidation of an endangerment finding. If a statute is precautionary in nature and designed to protect the public health, and the relevant evidence is difficult to come by, uncertain, or conflicting because it is on the frontiers of scientific knowledge, EPA need not provide rigorous step-by-step proof of cause and effect to support an endangerment finding. As we have stated before, [a]waiting certainty will often allow for only reactive, not preventive, regulation." Coalition for Responsible Regulation, Inc. v. EPA, 684 F.3d 102, 121 (D.C. Cir. 2012) (*internal quotations and citations omitted*).

[89] EPA Office of Research Development, National Center for Environmental Assessment, Global Change Research Program, *Climate Change Vulnerability Assessments: Four Case Studies of Water Utility Practices* (Washington, D.C.: 2011).

[90] NRC, America's Climate Choices: Panel on Adapting to the Impacts of Climate Change, *Adapting to the Impacts of Climate Change* (Washington, D.C.: 2010).

[91] For more information about NASA's Climate Change Adaptation Policy Statement, visit http://www.nasa.gov/agency/sustainability/.

[92] Additional information on the October 5, 2009, Executive Order 13514 on Federal Leadership in Environmental, Energy, and Economic Performance is available at http://www.whitehouse.gov/administration/eop/ceq/sustainability. For more information about the Climate Change Adaptation Task Force, see http://www.whitehouse.gov/administration/eop/ceq/initiatives/adaptation.

[93] OSTP was established by statute in 1976 to serve as a source of scientific and technology analysis judgment.

[94] This is not a comprehensive description of the activities of the Interagency Climate Change Adaptation Task Force. Task force activities relevant to adapting infrastructure to a changing climate—such as agency adaptation plans—are discussed in more detail in different sections of this report.

[95] NRC, Committee on *a National Strategy for Advancing Climate Modeling*, Board on Atmospheric Studies and Climate, Division on Earth and Life Sciences, A National Strategy for Advancing Climate Modeling (Washington, D.C.: 2012).

[96] For more information on the *National Fish, Wildlife, and Plants Climate Adaptation Strategy* and other strategic adaptation planning efforts, visit http://www.whitehouse.gov/administration/eop/ceq/initiatives/adaptation.

[97] Pub. L. No. 101-606, § 106 (1990).

[98] In 2007, a federal district court held that the failure to complete an assessment at least every 4 years violated the Global Change Research Act. Center for Biological Diversity v. Brennan, 571 F.Supp.2d 1105, 1131 (N.D. Cal. 2007).

[99] For more information on the third National Climate Assessment, visit http://www.globalchange.gov/what-we-do/assessment.

[100] In technical comments, CEQ, OSTP, and USGCRP noted that the third National Climate Assessment also includes an assessment of the state of adaptation and the state of decision support in the United States.

[101] Additional information on FHWA's climate change vulnerability assessment pilots can be found at http://www.fhwa.dot.gov/environment/climate_change/adaptation/ ongoing_and_current_research/vulnerability_assessment_pilots/index.cfm.

[102] Written comments from the Department of Transportation stated that the department has sought to develop vulnerability assessment tools, as well as risk assessment tools to help state and local agencies identify critical assets that are potentially vulnerable to climate change. These comments also noted that the Department of Transportation has developed and deployed a climate information resource to help state and local agencies better frame their potential risks.

[103] For more information on the Federal Highway Administration's December 2012 *Climate Change and Extreme Weather Vulnerability Assessment Framework*, visit http://www.fhwa.dot.gov/environment/climate_change/adaptation/resources_and_publications/ vulnerability_assessment_framework/index.cfm

[104] In written comments, the Department of Transportation stated that the Federal Highway Administration has been leading and advocating for awareness and advancement of adaptation issues. The Department of Transportation has developed a Framework and Strategic Vision to aide in the development of plans, stewardship, and oversight between it and the various state transportation agencies and has also convened several national and regional workshops and peer-exchanges with interested parties to begin and advance the dialog. Further, according to these comments, there is increasing recognition of climate change adaptation at least in part to the Federal Highway Administration's outreach efforts with states.

[105] NRC, America's Climate Choices: Panel on Informing Effective Decisions and Actions Related to Climate Change, *Informing an Effective Response to Climate Change* (Washington, D.C.: 2010).

[106] USGCRP, National Global Change Research Plan 2012-2021: *A Strategic Plan for the U.S. Global Change Research Program* (Washington D.C.: April 2012).

[107] National Drinking Water Advisory Council, *Climate Ready Water Utilities Final Report* (Dec. 9, 2010).

[108] USGCRP, National Global Change Research Plan 2012-2021: *A Strategic Plan for the U.S. Global Change Research Program* (Washington, D.C.: April 2012).

[109] GAO, *Climate Change: Improvements Needed to Clarify National Priorities and Better Align Them with Federal Funding Decisions*, GAO-11-317 (Washington, D.C.: May 20, 2011).

[110] NRC, *Restructuring Federal Climate Research to Meet the Challenges of Climate Change* (Washington, D.C.: 2009).

[111] NRC, America's Climate Choices: Panel on Informing Effective Decisions and Actions Related to Climate Change, *Informing an Effective Response to Climate Change* (Washington, D.C.: 2010).

[112] USGCRP, National Global Change Research Plan 2012-2021: *A Strategic Plan for the U.S. Global Change Research Program* (Washington, D.C.: April 2012).

[113] According to technical comments from CEQ, OSTP, and USGCRP, the Global Change Information System will include an adaptation clearinghouse if all goes as planned. Stage one of the Global Change Information System is deployment of the data supporting the third National Climate Assessment.

[114] Determining how to prioritize these activities is beyond the scope of this report.

[115] GAO

[116] For more information on EPA's Climate Resilience Evaluation and Awareness Tool, please visit http://water.epa.gov/infrastructure/watersecurity/climate/creat.cfm.

[117] The Department of Transportation's transportation and climate change clearinghouse is available at http://climate.dot.gov/. In technical comments, the Department of Transportation noted that its staff is currently working to overhaul the Climate Clearinghouse to better serve the needs of decision makers who frequent the site.

[118] U.S. Climate Change Science Program, *Impacts of Climate Change and Variability on Transportation Systems and Infrastructure*: Gulf Coast Study, Phase I. A Report by the U.S. Climate Change Science Program and the Subcommittee on Global Change Research (Savonis, M. J., V.R. Burkett, and J.R. Potter [eds.]). Department of Transportation (Washington, D.C.: 2008).

[119] For more information on Phase 2 of the Gulf Coast Study, visit http://www.fhwa.dot.gov/environment/climate_change/adaptation/ongoing_and_current_research/gulf_coast_study/index.cfm

[120] USGCRP, National Global Change Research Plan 2012-2021: *A Strategic Plan for the U.S. Global Change Research Program* (Washington, D.C.: April 2012).

[121] NRC, Panel on Strategies and Methods for Climate-Related Decision Support, Committee on the Human Dimensions of Global Change, *Informing Decisions in a Changing Climate* (Washington, D.C.: 2009).

[122] The University of Washington Climate Impacts Group is no longer funded as a NOAA RISA project. NOAA's RISA for the Pacific Northwest is now the Climate Impacts Research Consortium, a consortium of three multiuniversity organizations.

[123] NRC, Panel on Strategies and Methods for Climate-Related Decision Support, Committee on the Human Dimensions of Global Change, *Informing Decisions in a Changing Climate* (Washington, D.C.: 2009).

[124] Pub. L. No. 91-190 (1970), codified as amended at 42 U.S.C. §§ 4321-4347 (2011). Under NEPA, federal agencies must assess the effects of major federal actions—those they propose to carry out or to permit—that significantly affect the environment. NEPA has two principal purposes: (1) to ensure that an agency carefully considers detailed information concerning significant environmental impacts and (2) to ensure that this information will be made available to the public.

[125] In technical comments, the Department of Transportation noted that it is probably more accurate to say that federal agencies rely on professional associations in adopting design standards rather than saying that federal agencies do not develop design standards. The point, according to the department's technical comments, is that professional associations develop the design standards in general that are required to be used in their programs, not just design standards for climate change.

[126] According to EPA officials, EPA programs generally specify performance standards— rather than design standards—for infrastructure to protect water quality. This, in turn, drives engineers' selection of infrastructure design based on a variety of factors.

[127] Resources for the Future, *Summary Report. Reforming Institutions and Managing Extremes: U.S. Policy Approaches for Adapting to a Changing Climate* (Washington D.C.: 2011).

[128] For more information about AASHTO's Transportation and Climate Change Resource Center, see http://climatechange.transportation

[129] In technical comments, CEQ, OSTP, and USGCRP noted that in at least some cases, a more active federal role in considering climate change information in design standards would require statutory authorization or executive action.

[130] NRC, America's Climate Choices: Panel on Adapting to the Impacts of Climate Change, *Adapting to the Impacts of Climate Change* (Washington, D.C.: 2010).

[131] S. 1813, 112th Cong. (2011).

[132] If, however, the agency determines that activities of a proposed project fall within a category of activities the agency has already determined has no significant environmental impact—called a categorical exclusion—then the agency generally need not prepare an environmental assessment or environmental impact statement.

[133] CEQ coordinates federal environmental efforts and works closely with agencies and other White House offices in the development of environmental policies and initiatives. CEQ was established within the Executive Office of the President by the National Environmental Policy Act of 1969, and its role was expanded by the Environmental Quality Improvement Act of 1970.

[134] According to technical comments from CEQ, OSTP, and USGCRP submitted to GAO on March 22, 2013, CEQ is still in the process of developing NEPA guidance on greenhouse gases and climate change.

[135] For more information about agency adaptation plans, visit http://sustainability.performance.gov/. According to technical comments from the Department of Transportation, the department's adaptation plan, among other things, notes that the Federal Highway Administration will develop in fiscal years 2012 and 2013 draft guidance documenting procedures and methodologies for incorporating climate change considerations into planning and design analyses for highway projects in the coastal environment. More specifically, the plan states that the Federal Highway Administration will provide information on the state of the practice for addressing climate change in analyses related to sea level rise, storm surge, and wave action and that the results will be used to support transportation decision making. According to these technical comments, subsequent updates to the Department of Transportation's adaptation plan will reflect consideration of any comments received on the first plan that was released in February 2013.

[136] We have additional climate change adaptation work under way. For example, we expect to release a report on climate change adaptation and natural resource planning in mid- 2013. We also expect to complete work on energy and water infrastructure adaptation later in 2013.

[137] For more information about the Climate Adaptation Knowledge Exchange, visit http://www.cakex.org/.

[138] For more information about the Georgetown Climate Center, visit http://www.georgetownclimate.org/.

In: Climate Change and Infrastructure
Editor: Dustin Ethans
ISBN: 978-1-63463-061-0

Chapter 2

CLIMATE CHANGE SCIENCE: KEY POINTS*

Jane A. Leggett

SUMMARY

Though climate change science often is portrayed as controversial, broad scientific agreement exists on many points:

- The Earth's climate is warming and changing.
- Human-related emissions of greenhouse gases (GHG) and other pollutants have contributed to warming observed since the 1970s and, if continued, would tend to drive further warming, sea level rise, and other climate shifts.
- Volcanoes, the Earth's relationship to the Sun, solar cycles, and land cover change may be more influential on climate shifts than rising GHG concentrations on other time and geographic scales. Human-induced changes are super-imposed on and interact with natural climate variability.
- The largest uncertainties in climate projections surround feedbacks in the Earth system that augment or dampen the initial influence, or affect the pattern of changes. Feedback mechanisms are apparent in clouds, vegetation, oceans, and potential emissions from soils.

---

* This is an edited, reformatted and augmented version of a Congressional Research Service publication, CRS Report for Congress R43229, prepared for Members and Committees of Congress, from www.crs.gov, dated September 10, 2013.

- There is a wide range of projections of future, human-induced climate change, all pointing toward warming and associated sea level rise, with wider uncertainties regarding the nature of precipitation, storms, and other important aspects of climate.
- Human societies and ecosystems are sensitive to climate. Some past climate changes benefited civilizations; others contributed to the demise of some societies. Small future changes may bring benefits for some and adverse effects to others. Large climate changes would be increasingly adverse for a widening scope of populations and ecosystems.

As is common and constructive in science, scientists debate finer points. For example, a large majority but not all scientists find compelling evidence that rising GHG have contributed the most influence on global warming since the 1970s, with solar radiation a smaller influence on that time scale. Most climate modelers project important impacts of unabated GHG emissions, with low likelihoods of catastrophic impacts over this century. Human influences on climate change would continue for centuries after atmospheric concentrations of GHG are stabilized, as the accumulated gases continue to exert effects and as the Earth's systems seek to equilibrate.

The U.S. government and others have invested billions of dollars in research to improve understanding of the Earth's climate system, resulting in major improvements in understanding while major uncertainties remain. However, it is fundamental to the scientific method that science does not provide absolute proofs; all scientific theories are to some degree provisional and may be rejected or modified based on new evidence. Private and public decisions to act or not to act, to reduce the human contribution to climate change or to prepare for future changes, will be made in the context of accumulating evidence (or lack of evidence), accumulating GHG concentrations, ongoing debate about risks, and other considerations (e.g., economics and distributional effects).

Broad Scientific Agreement on Many Aspects of Climate Change[1]

Despite portrayals in popular media about controversies in climate change science, almost all climate scientists agree on certain important points:

- The Earth's climate has warmed significantly and changed in other ways over the past century (*Figure 1*). The warming has been widespread but not uniform globally, with most warming over continents at high latitudes, and slight cooling in a few regions, including the southeastern United States, the Amazon, and the North Atlantic south of Greenland.[2]

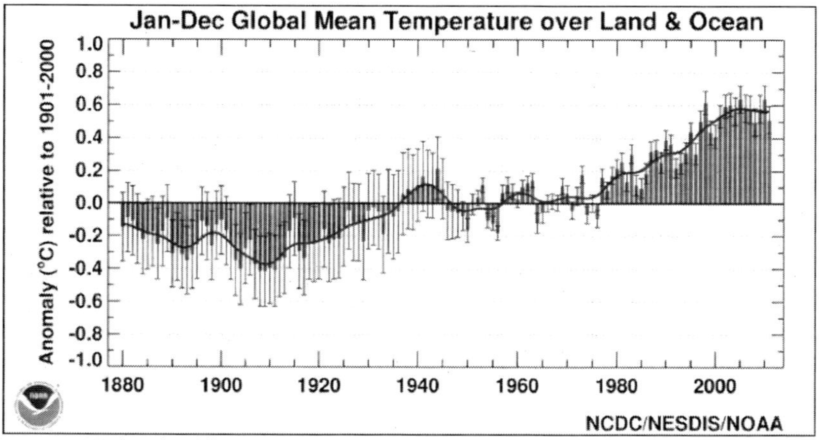

Source: National Climate Data Center, National Oceanic and Atmospheric Administration, U.S. Department of Commerce. Figure extracted March 28, 2013. Very similar findings are reported by several other, independent research groups. See, for example, Rohde, Robert, Richard Muller, Robert Jacobsen, Elizabeth Muller, Saul Perlmutter, Arthur Rosenfeld, Jonathan Wurtele, Donald Groom, and Charlotte Wickham. "A New Estimate of the Average Earth Surface Land Temperature Spanning 1753 to 2011." Geoinformatics & Geostatistics: An Overview 1, no. 1 (December 7, 2012).

Notes: Red bars represent "anomalies," or differences in mean temperature for the year compared with the 20th century average. Anomalies are a better estimate than the absolute value, as they can capture the change over time more reliably while absolute values are vulnerable to gaps in geographic coverage. The blue line shows the running average, applying a "21-point binomial filter" to the time series plotted as red bars. The "whisker" (thin black vertical) lines represent confidence or possible error levels. Confidence has improved over the past century.

Figure 1. Long-Term Temperature Observations. Compared to the 20th Century Global Mean Temperature.

- The climate has varied naturally through geologic history. Past climate changes sometimes proceeded abruptly when they passed certain "tipping points." The National Academy of Sciences concluded that the past few decades were very likely the warmest in the past 400 years, and "that temperatures at many, but not all, individual locations were higher during the past 25 years than during any period of comparable length since A.D. 900."[3] Although conclusions cannot yet be precise, research suggests that global average temperatures today are among the highest since human civilizations began to flourish roughly 4,000 to 8,000 years ago.[4]

- "Greenhouse gases" (GHG) include, among others, carbon dioxide (CO_2), water vapor, methane (CH_4), and nitrous oxide (N_2O), as well as some aerosols. They absorb energy into the atmosphere rather than letting it escape to space. The presence of GHG in the atmosphere warms the Earth to its current temperature.
- Human activities, especially fossil fuel burning, deforestation, agriculture, and some types of industry, have increased GHG concentrations in the atmosphere. CO_2, the main GHG emitted by human activities, has risen almost 40% over the past 150 years. About one-third of human-related CO_2 has been absorbed by oceans, increasing surface water acidity by 30%.[5]
- The enhanced levels of GHG in the atmosphere contributed to the observed global average warming in recent decades. Over other time and geographic scales, such factors as the Earth's orbit, solar variability, volcanoes, and land cover change have been stronger influences than human-related GHG.
- There is a wide range of projections of future human-induced climate change, with all pointing toward warming. Human-induced change will be superimposed on, and interact with, natural climate variability.
- Human societies and ecosystems are sensitive to climate. Some climate changes benefited civilizations; others contributed to some societies' demises.
- The range of possible impacts on humans and ecosystems is also very wide. In the near term, climate change (including the fertilization of vegetation by CO_2) may bring benefits for some, while adversely affecting others. Researchers expect the balance of projected climate change impacts to be increasingly adverse for a widening scope of populations and ecosystems.

As is common and constructive in science, scientists debate finer points. Some disagree with the broader consensus that GHG have been *the major* influence on global warming over the past few decades. Some suggest that, if GHG emissions continue unabated, the resulting climate change would be small and possibly beneficial overall. Most climate modelers project changes that are significant to large, with small likelihoods of changes that could be catastrophic for some human societies and ecosystems in coming decades.

DEALING WITH UNCERTAINTIES

Even the best science cannot provide absolute proof; it is fundamental to the scientific method that all theories are to some degree provisional and may be rejected or modified based on new evidence. Private and public decisions to act or not to act, to reduce the human contribution to climate change or to prepare for future changes, will be made in the context of accumulating evidence (or lack of evidence), accumulating GHG concentrations, ongoing debate about risks, and other considerations (e.g., economics and distributional effects). That said, billions of dollars have been invested in research on a wide range of climate change topics, including the possibility of attribution to alternative causes than greenhouse gases. To date, scientists have found little support for the hypothesis that GHG are not responsible for observed warming, nor have they found much evidence that other factors (including solar changes) can explain more than a small portion of global average temperature increases since the 1970s. For example, measurements of solar irradiance suggest that the solar influence on global temperatures has been decreasing overall since the 1930s, with the up-and-down pattern of the 11-year solar cycle evident in observations. A large body of research is consistent with attributing the majority of global temperature increase since the 1970s to the increase in GHG concentrations. It is this balance of evidence that leads most scientists to consider human-related GHG emissions an important global risk.

Sound Science Does Not Offer Proof

As scientists may point out, "there is no such thing as a scientific proof. Proofs exist only in mathematics and logic, not in science.... The primary criterion and standard of evaluation of scientific theory is evidence, not proof.... The currently accepted theory of a phenomenon is simply the best explanation for it among all available alternatives."6 Normal scientific methods aim at disproving a hypothesis; if evidence cannot disprove a hypothesis, it generally buttresses confidence in the hypothesis.

ISSUES FOR CONGRESS

It appears unlikely that science will provide decision-makers with significantly more scientific certainty for many years regarding the precise patterns and risks of climate change. Nonetheless, both private and public decision-makers face climate-related choices.

Broadly, response options to significant climate change include (1) defer the choices; (2) find out more; (3) inform affected populations; (4) prepare; (5) try to contain it; and (6) choose to experience the consequences. In many cases, many decision-makers are likely to face situations that require a response, such as resolving discrepancies between designated and actual flood plains or attempting to improve agricultural productivity in light of contemporary climate patterns.

Based on what is and what is not well known concerning climate change, as well as other considerations, Members of Congress may address climate-related decisions that affect

- authorizations and appropriations for federal programs, including research and technology development;
- tax and financial incentives for private decision-makers;
- regulatory authorities; or
- information or assistance to affected entities to help them adapt or rebuild after damages.

A variety of other CRS reports provide background and analysis on such options and are listed at the end of this report.[7]

Causes of Observed Climate Change: Forcings, Feedbacks, and Internal Variability

Three concepts may be useful for understanding the mechanisms and debate over the contributions to observed climate change: *forcings, feedbacks,* and *internal variability.*

Forcings

There is broad agreement among scientists that certain factors—including the composition of the atmosphere and solar variability—directly change the balance between incoming and outgoing radiation in the Earth's system and consequently *force* climate change. *Forcings* include the following:

- Atmospheric concentrations of greenhouse gas (GHG) and other trace gas and aerosol. These include water vapor,8 carbon dioxide (CO_2), methane (CH_4), nitrous oxide (N_2O), sulfur hexaflouride (SF6), chlorofluorocarbons (CFC), hydrofluorocarbons (HFC), perfluorocarbons (PFC), ozone, sulfur aerosols, black and organic carbon aerosols, and others. Human activities, especially fossil fuel burning, deforestation, agriculture, and some types of industry, have increased GHG concentrations in the atmosphere. CO_2 has risen almost 40% over the past 150 years.[9]
- Amount and patterns of solar radiation reaching the Earth due to the Earth's orbit around the Sun, and the tilt and wobble of the Earth's axis, as well as solar variability (Figure 2).
- Reflectivity of the Earth's surface due to changes in land use (e.g., urban surfaces, forest cover), changes in ice and snow cover; and vegetation cover.

Human-Related Greenhouse Gas Emissions

A majority of human-related GHG emissions are carbon dioxide, released primarily from energy production and use, deforestation and forest degradation, and cement manufacture. World-wide in 2010, carbon dioxide emissions were 74% of human-related GHG emissions. In the United States, carbon dioxide was 83% of human-related GHG emissions (*Figure 3*). Methane and nitrous oxide emissions are greater shares (16% and 8%, respectively) globally than in the United States (10% and 5%, respectively). Agriculture is a main source of these emissions and is a bigger share of the economies of many low-income countries compared with the United States. Also, many sources in the United States have acted to reduce their GHG emissions (such as in reducing leaks of methane), compared with sources in some low-income countries.

A large majority (73%) of global GHG are emitted by the 10 top emitting countries *Figure 4*). China became the leading GHG emitter in 2007 when it surpassed the United States. While China's emissions have been on the rise, the United States has emitted more cumulatively than any other country over the past 100 years.

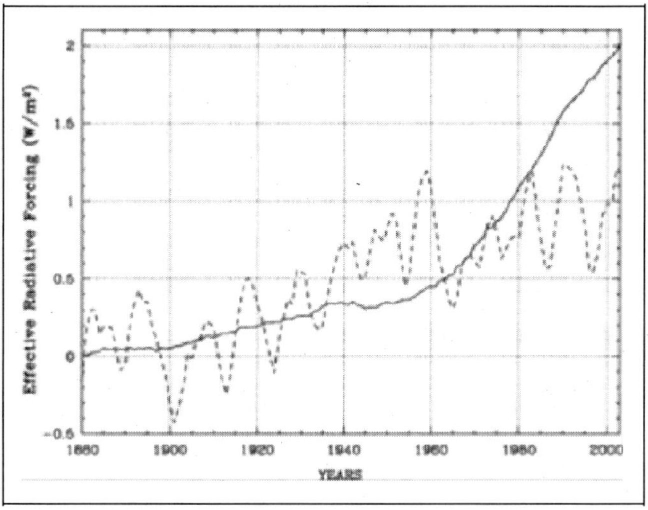

Source: Ziskin, Shlomi, and Nir J. Shaviv. "Quantifying the Role of Solar Radiative Forcing over the 20th Century." Advances in Space Research 50, no. 6 (September 15, 2012): 762–776. doi:10.1016/j.asr.2011.10.009.

Notes: According to the authors, "the optimal anthropogenic contribution (solid line) and the optimal solar contribution (dashed line) over the 20th century. The anthropogenic contribution is primarily composed of GHGs and aerosols. The solar contribution includes changes in the total solar irradiance and the indirect solar effect (ISE)." This is one of many studies, using a variety of methods, investigating the relative contributions of different climate forcings that conclude that the GHG concentrations have outweighed all other influences on global mean air surface temperature from the late 1970s to the present. For a broader, more thorough review of scientific understanding of the solar influence, see Gray, L. J., J. Beer, M. Geller, J. D. Haigh, M. Lockwood, K. Matthes, U. Cubasch, et al. "Solar Influences on Climate." Reviews of Geophysics 48, no. 4 (October 30, 2010). doi:10.1029/2009RG000282.

Figure 2. One Estimate of Human-Related versus Solar Contributions to Global Temperature Change Over the 20th Century.

Feedbacks

Once a change in the Earth's climate system is underway, responses *within* the system will amplify or dampen the initiated change. Virtually all climate scientists conclude that all the feedbacks *in net* are likely to be positive (i.e., increasing climate change in the same direction caused by warming),[10] especially if temperature increases are large;[11] there remain wide differences in views.

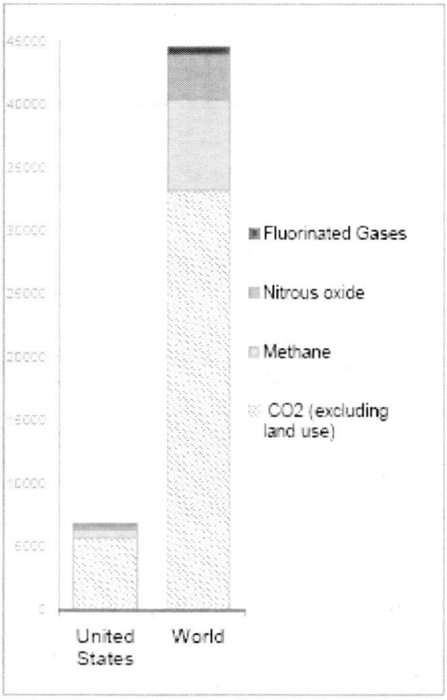

Source: CRS figure using estimates from World Resources Institute, CAIT version 2.0, extracted September 12, 2013.

Notes: These estimates cover the six GHG covered by the Kyoto Protocol (CO_2, CH_4, N_2O, SF6, HFC, and PFC), expressed in their equivalencies to the effect of CO_2 on "radiative forcing" of the atmosphere over a 100-year period.

The World column includes U.S. emissions.

Figure 3. Shares of Human-Related GHG Emissions by Gas in 2010. Million metric tons of CO_2-equivalent.

An important consideration is that, once positive feedbacks begin, they may be essentially irreversible and, at least theoretically, lead to "runaway warming." A few of the major feedbacks are clouds, vegetation, snow and ice cover, and uptake or releases of GHG by soils and water bodies. Forests, for example, provide both negative and positive feedbacks. On the one hand, higher CO2 concentrations in the atmosphere tend to fertilize their growth (if other conditions are not limiting) and forests may grow more rapidly with greater warmth and precipitation; these factors could dampen initiated warming. On the other hand, forests thrive within certain bounds of growing conditions; if their climate conditions change beyond those bounds, they are likely to grow more slowly and eventually die back, releasing the carbon they and forest soils store and enhancing the initiated climate change.

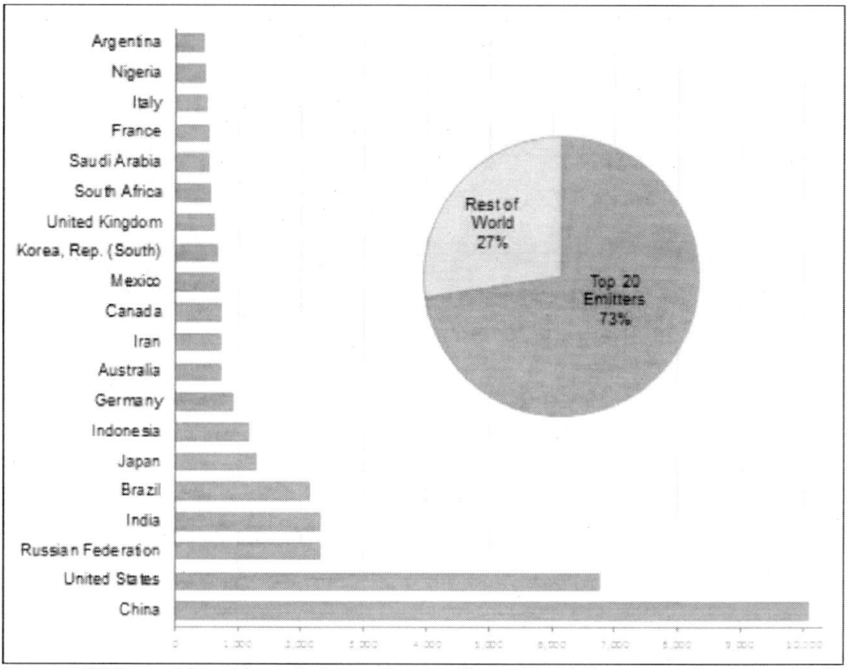

Source: CRS graphic with emission estimates from World Resources Institute, CAIT Version 2.0, extracted September 12, 2013.

Notes: These estimates cover the six GHG covered by the Kyoto Protocol (CO_2, CH_4, N_2O, SF6, HFC, and PFC), expressed in their equivalencies to the effect of CO_2 on "radiative forcing" of the atmosphere over a 100- year period.

Figure 4. Estimated Top 20 Emitting Nations of Greenhouse Gases in 2010. Million metric tons of carbon-dioxide equivalent, includes all net land use fluxes.

Internal Variability

The climate exhibits its own rhythms, or *internal variability*. The oscillation between El Niño and La Niña events is an example of internal climate variability that has important effects on economies and ecosystems in the Pacific basin (including across the United States). Another is the North Atlantic Oscillation. Internal variability may be difficult to distinguish from decadalscale climate change. Such patterns of variability also may be influenced by climate change.

PROJECTIONS OF FUTURE HUMAN-INDUCED CLIMATE CHANGE

Most climate science experts project that if GHG emissions are not reduced far below current levels, the Earth's climate would warm further, above natural variations, to levels never experienced by human civilizations. If, and as, the climate moves further from its present state, it would reconfigure the patterns and events to which current human and ecological systems are adapted, and the risk of abrupt changes would dramatically increase.

Scenarios of future GHG concentrations under current policies range from 500 ppm carbon dioxide equivalents[12] (CO_2e) to over 1,000 ppm CO_2e by 2100. These are projected to raise the global average temperature by at least 1.5° Celsius (2.7° Fahrenheit) above 1990 levels,[13] not taking into account natural variability. The estimates considered most likely by many scientists are for GHG-induced temperature increases around 2.5 to 3.2° C (4.5 to 5.8° F) by 2100.[14] There is a small but not trivial likelihood that the GHG-induced temperature rise may exceed 6.4°C (11.5° F) above natural variability by 2100.[15]

As context, the global average temperature at the Last Glacial Maximum has been estimated to be about 3 to 5° C (5.4 to 9° F) cooler than present,[16] and is estimated currently to be approaching the highest level experienced since the emergence of human civilizations about 8,000 years ago.[17]

Future climate change may advance relatively smoothly or sporadically, and some regions are likely to experience more fluctuations in temperature, precipitation, and frequency or intensity of extreme events than others. Almost all regions are expected to experience warming; some are projected to become warmer and wetter, while others would become warmer and drier. Sea levels could rise due to ocean warming alone on average between 7 and 23 inches by 2100. Adding to that estimate would be the effects of poorly understood but possible accelerated melting of the Greenland or Antarctic ice sheets. Recent scientific studies have projected a total global average sea level to rise in the 21st century, depending on GHG scenarios, ice dynamics, and other factors, in the range of 2 to 2.5 feet, with a few estimates ranging up to 6.5 feet.[18] Continued warming could lead to additional sea level rise over subsequent centuries of several to many meters. Improving understanding of ice dynamics is a high priority for scientific research to improve sea level rise projections.

Patterns consistent among different climate change models have led to some common expectations: GHG-induced climate change would include more heat waves and fewer extreme cold episodes; more precipitation on average but more droughts in some regions; and generally increased summer warming and dryness in the central portions of continents. Regional changes may vary from the global average changes, however. Scientists also expect precipitation to become more intense when it occurs, thereby increasing runoff and flooding risks.

Precipitation is a particularly challenging component of projecting future climate. For example, for the contiguous United States, recent climate modeling consistently anticipates overall temperature increases, but different models produce a wide range of precipitation changes, from net decreases to net increases.[19] This is particularly problematic in that precipitation, and its characteristics, is closely associated with impacts on agriculture, water supply, streamflows, and other critical systems.

Scientific expectations and model projections consistently point to a global average increase in precipitation with strong variations across regions and time. Generally, dry areas are expected to get dryer, and wet regions are expected to get wetter. In many regions, the increase in evapotranspiration is expected to exceed the increase in precipitation, resulting in general drying of soils and increasing risks of droughts. Precipitation, when it occurs, is expected to be more intense. There will be more energy available for storms, including hurricanes and thunderstorms, though whether they may increase in frequency remains unclear. Sea ice cover in the Arctic is projected to continue its recent decline (*Figure 5*). Greenland is expected to continue ice loss, adding to sea level rise, with more uncertainty about what may happen to ice cover in Antarctica (*Figure 5*). Because Arctic sea ice already floats on water, its melting would not increase sea levels, but large scale melting of land-based ice in Greenland and Antarctica could increase average sea levels by as many as 2 meters by 2100 and several more meters over coming centuries.

IMPACTS OF CLIMATE CHANGE

Nearly every human and natural system could be affected by climate changes, directly or indirectly. The U.S. Global Change Research Program has produced several assessments of scientific understanding of impacts of climate change on the United States.[20]

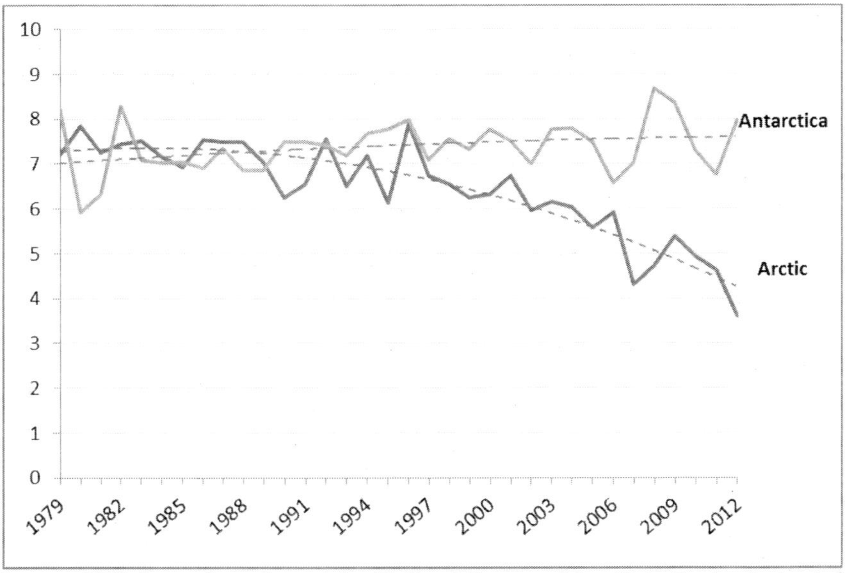

Source: CRS figure from data at the National Snow and Ice Data Center (extracted March 29, 2013), at http://nsidc.org/data/docs/noaa/g02135_seaice_index/ # monthly_graphs_format.

Notes: The dotted lines show the best polynomial fit (2-order) to each time series, as estimated by Excel. For both series, the polynomical fit was slightly better than a linear fit. See the National Snow and Ice Data Center website, referenced above, for further description of the underlying data.

Figure 5. Sea Ice Extent in Arctic (September) and Antarctica (April), 1979 to 2012. (in millions of square kilometers)

Climate Changes Would Affect A Wide Set of Human Systems

Changes in patterns of temperature, precipitation, sea levels, storms, and heat waves (among other indicators of climate) would affect, among other systems:

- water resources and delivery;
- agricultural productivity;
- the frequency and intensity of extreme weather events;
- spread of infectious diseases; air and water pollution levels;
- reliability of transportation, energy, and coastal protection systems;

- commodity prices;
- insurance pay-outs; and
- migration of people and species.

There are many additional elements of the economy and society that could be affected by shifts in climate. Research on potential impacts of climate change is generally less funded and developed than on the climate system itself.

Whether climate changes are meaningful in a policy context arguably depends, on the one hand, on how they influence existing and emerging human systems, and on the other hand, the values people attach to different resources and risks. Past climate changes, often regional not global, contributed to major societal changes, including some large-scale migrations and even the demise of some civilizations.[21] Some climate changes likely stimulated technological advances, such as development of irrigation systems.

Many investments in current buildings, transportation, water systems, agricultural hybrid varieties, and other infrastructure were designed in the context of a climate of one or more decades ago, cooler on average than today. To the degree that climate patterns were factored into design, the investments typically presumed that climate would remain stable within historical bounds of variability. For example, levees may have been built to withstand a 100-year flood (1% chance to occur each year) according to historic runoff, streamflow, and storm surge conditions going back many decades. As climate changes produce greater, more intense precipitation and run-off, however, a 100-year flood may now be closer to the 50-year flood (2% chance flood), and potentially the 10-year flood within decades (10% change flood). If climate continues to change from the conditions for which infrastructure and practices were designed, the risks of losses due to maladaptation would increase.

A wide band of uncertainty surrounds projections of impacts of climate change and, in particular, the critical thresholds for non-linear or abrupt effects. Some impacts of climate change are expected to be beneficial in some locations with a few degrees of warming (e.g., increased agricultural productivity in some regions, less need for space heating, less cold weather mortality, opening of the Northwest Passage for shipping and resource exploitation). Most impacts are expected to be adverse (e.g., lower agricultural productivity in many regions, drought, rising sea levels, spread of disease vectors, greater needs for cooling). Many impacts may be substantial but hard to assess as yet as "positive" or "negative," such as possible impacts on the structure of global financial markets. Risks of abrupt, surprising climate changes with accompanying dislocations are expected to increase as global average temperature increases; some could push natural and socioeconomic systems past key thresholds of tolerance.

Risks of future climate change would be reduced by efforts that reduce vulnerability and build resilience ("adaptation"). Some populations will have the resources to migrate and adapt successfully—even profit from new opportunities that will emerge—while others could lose livelihoods or lives. Adaptations can help mitigate impacts and damage costs, but also impose costs, often on those who can least afford them. Climate change will occur with different magnitudes and characteristics in different regions. The difficulties involved in improving predictions at regional and local scales will challenge preparations for climate change. To a large degree, climate change will expand the uncertainties that individuals and organizations face.

Climate change could have a wide array of effects on individuals, communities, and populations on a large scale. Many of these are expected to occur in small increments: shortages and increasing prices for clean water, rising food prices, higher rates of allergies and such illnesses as diarrhea or cholera, erosion of beaches, etc.

At an increasing rate may be shocks, or distinct weather events, such as more extreme heat waves, severe droughts, or loss of industrial cooling systems when intake water is in short supply or is warmer than tolerable temperatures.[22]

Atmospheric Carbon Dioxide is Increasing Ocean Acidity

The *acidity of the surface waters of the oceans* has increased by about 26% over the past 150 years.[23] Ocean acidification has occurred along with the rise in atmospheric concentrations of CO_2. The oceans remove 25%-40% of the carbon dioxide emissions added annually to the atmosphere by burning fossil fuels. The carbon dioxide absorbed in the oceans decreases the water's pH, an indicator of increasing acidity. According to a National Research Council (NRC) report, the current rate of acidification "exceeds any known change in ocean chemistry for at least 800,000 years."[24] Research shows varying sensitivities of different marine species to rising acidity, making general statements about impacts of ocean acidification difficult.

The NRC concluded,

> While the ultimate consequences are still unknown, there is a risk of ecosystem changes that threaten coral reefs, fisheries, protected species, and other natural resources of value to society. (*Executive Summary*, pp. 3-4)
>
> Congress enacted the Federal Ocean Acidification Research and Monitoring Act of 2009 (P.L. 111-11, Section 12311, Subtitle D) to improve monitoring and research, to assess carbon storage in the oceans and potential effects on acidification and other ocean conditions, and to develop predictive models for future changes in ocean chemistry and marine ecosystems. The program is housed within the National Oceanic and Atmospheric Administration (NOAA), and coordinated with other agencies through an interagency plan through the National Ocean Council.

Extreme events, chronic economic losses, or improved opportunities elsewhere are expected to prompt migration of millions of people, largely within countries, but also across national borders. Extreme events, greater variability, and uncertainty are expected to increase stress and mental health challenges. Some experts project that climate changes could amplify instabilities in countries with weak governance and increase security risks.[25] This may have implications for international political stability and security.[26]

For some experts and stakeholders, likely ecological disruptions (and limitations on species' and habitats' abilities to adapt at the projected rate of climate change) are among the most compelling reasons that humans must act to reduce their interference with the climate system. Some believe humans will have the wherewithal to cope, but non-human systems may not. As the degree and distribution of climate changes continue, ranges of species are likely to change. Climate change is highly likely to create substantial changes in ecological systems and services[27] in some locations, and may lead to ecological surprises.[28] The disappearance of some types of regional ecosystems raises risks of extinctions of species, especially those with narrow geographic or climatic distributions, and where existing ecological communities disintegrate.[29] One set of researchers found "a close correspondence between regions with globally disappearing climates and previously identified biodiversity hotspots; for these regions, standard conservation solutions (e.g., assisted migration and networked reserves) may be insufficient to preserve biodiversity."[30]

SELECTED, RELATED CRS REPORTS

CRS Report R43185, *Ocean Acidification*, by Harold F. Upton and Peter Folger.

CRS Report R41153, *Changes in the Arctic: Background and Issues for Congress*, coordinated by Ronald O'Rourke.

CRS Report RL34580, *Drought in the United States: Causes and Issues for Congress*, by Peter Folger, Betsy A. Cody, and Nicole T. Carter.

CRS Report R43199, *Energy-Water Nexus: The Energy Sector's Water Use*, by Nicole T. Carter.

CRS Report R42611, *Oil Sands and the Keystone XL Pipeline: Background and Selected Environmental Issues*, coordinated by Jonathan L. Ramseur.

CRS Report R42756, *Energy Policy: 113th Congress Issues*, by Carl E. Behrens.

CRS Report R42613, *Climate Change and Existing Law: A Survey of Legal Issues Past, Present, and Future*, by Robert Meltz.

CRS Report R43120, *President Obama's Climate Action Plan*, coordinated by Jane A. Leggett. CRS Report R42756, *Energy Policy: 113th Congress Issues*, by Carl E. Behrens. CRS Report RL34266, *Climate Change: Science Highlights*, by Jane A. Leggett.

CRS Report R41973, *Climate Change: Conceptual Approaches and Policy Tools*, by Jane A. Leggett.

End Notes

[1] This CRS report will be reviewed and, as appropriate, revised considering evidence provided from emerging scientific research. Of, note, the Intergovernmental Panel on Climate Change (IPCC) will release its fifth assessment report later in 2013.

[2] For more information, see maps available at the National Climate Data Center, http://www.ncdc.noaa.gov/oa/climate/globalwarming.html and http://www. ncdc.noaa.gov /oa/climate/research/trends.html#global.

[3] Board on Atmospheric Sciences and Climate. Surface Temperature Reconstructions for the Last 2,000 Years. National Research Council, 2006. http://books.nap.edu/openbook.php? record_id=11676&page=1.

[4] Marcott, Shaun A., Jeremy D. Shakun, Peter U. Clark, and Alan C. Mix. "A Reconstruction of Regional and Global Temperature for the Past 11,300 Years." Science 339, no. 6124 (March 8, 2013): 1198–1201. doi:10.1126/science.1228026; Kellerhals, T., S. Brütsch, M. Sigl, S. Knüsel, H. W. Gäggeler, and M. Schwikowski. "Ammonium Concentration in Ice Cores: A New Proxy for Regional Temperature Reconstruction?" Journal of Geophysical Research: Atmospheres 115, no. D16 (2010): n/a–n/a. doi:10.1029/2009JD012603; Thibodeau, Benoît, Anne de Vernal, Claude Hillaire-Marcel, and Alfonso Mucci.

"Twentieth Century Warming in Deep Waters of the Gulf of St. Lawrence: A Unique Feature of the Last Millennium." Geophysical Research Letters 37, no. 17 (2010): n/a–n/a. doi:10.1029/2010GL044771. See also the references at http://www.globalwarmingart.com/wiki/File:Holocene_Temperature_Variations_Rev_png, which depict a collection of major temperature reconstructions of the Holocene, as well as the broad range of uncertainty of available estimates and the average of those estimates.

[5] National Research Council. Ocean Acidification: A National Strategy to Meet the Challenges of a Changing Ocean. Washington DC, 2013; Feely, Richard A. 2010. A Rational Discussion of Climate Change: The Science, the Evidence, the Response. Testimony before the House Committee on Science and Technology, Subcommittee on Energy and Commerce. Washington DC. (p.130). See also CRS Report R43185, Ocean Acidification, by Harold F. Upton and Peter Folger.

[6] See, for example, the discussion in Kanazawa, Satoshi. "Common misconceptions about science I: "Scientific proof"." Psychology Today, November 16, 2008. http://www.psychologytoday.com/blog/the-scientific-fundamentalist/ 200811/common-misconceptions-about-science-i-scientific-proof.

[7] Many CRS reports related to climate change may be found at Issues Before Congress: Climate Change Science, Technology, and Policy, at http://www.crs.gov/pages/subissue.aspx?cliid=3878&parentid=2522&preview=False.

[8] Water vapor is the most important GHG in the atmosphere but is understood not to be directly influenced by humans; it would be, however, involved in feedback mechanisms, discussed later.

[9] About one-third of human-related CO2 has been absorbed by oceans, increasing surface water acidity by 30%. See National Research Council. Ocean Acidification: A National Strategy to Meet the Challenges of a Changing Ocean. Washington, DC, 2013; Feely, Richard A. 2010. A Rational Discussion of Climate Change: The Science, the Evidence, the Response. Testimony before the House Committee on Science and Technology, Subcommittee on Energy and Commerce. Washington, DC. (p.130).

[10] One line of evidence is that carbon dioxide levels have varied closely with the Earth's temperature in and out of glacial periods over the past million years. These cycles are mostly triggered by changes in the Earth's orbit, tilt, and wobble. In some of these cycles, temperatures rose in advance of rising atmospheric carbon dioxide concentrations. Scientists generally interpret this as a tendency for positive climate warming feedbacks that increase carbon dioxide concentrations which then enhance warming, etc.—that the net positive feedbacks amplify an initial climate warming.

[11] Positive feedbacks could increase if and when, for example, large tracts of forests die back as a response to exceeding their climate thresholds of tolerance, or current permafrost thaws and releases the carbon it contains, or if reservoirs of methane hydrates destabilize.

[12] In order to show multiple gases of different potencies on a single scale, GHG have been indexed relative to the effect that a mass of CO2 would have over several time periods (because GHG remain in the atmosphere for different lengths of time, from days to tens of thousands of years). The index used for these estimates uses a 100-year time horizon, the most frequently used period.

[13] Intergovernmental Panel on Climate Change Working Group I, Climate Change 2007: The Physical Basis (Cambridge, UK: Cambridge University Press, 2007).

[14] As a point of reference, the global mean annual temperature during the 20th century is estimated to have been approximately 13.9o Celsius (57.0o Fahrenheit), according to NOAA's National Climate Data Center.

[15] Ibid.

[16] Intergovernmental Panel on Climate Change Working Group I. Climate Change 2007: The Physical Basis. Cambridge, UK: Cambridge University Press, 2007. Executive Summary.

[17] Highest temperatures of the Holocene may have occurred in one or more periods some 5,000 to 8,000 years ago, although sufficient data are not available for all parts of the globe to have reliable estimates of average global temperature. The oldest cities discovered date from approximately the same period, such as the extensive settlement of Byblos in present-day Lebanon, by about 6,000 years ago, or Medinat Al-Fayoum in Egypt, about 6,000 years old. Since the early to mid-Holocene, however, average temperatures appear to have been declining slowly, with notable periods of warming and cooling. The changes entailed in Holocene climate variability have been significant in terms of effects on humans and ecosystems, and have led to both benefits to, and the demise of, numerous civilizations.

[18] See discussion in National Research Council. Advancing the Science of Climate Change. Washington DC, 2010, at p. 244.

[19] See, for example, climate change scenarios available from the U.S. Global Change Research Program at http://scenarios.globalchange.gov/sites/default/ files/b/figures/ UnitedStates/ Ann_US_precip_a2.png, with notes at http://scenarios.globalchange.gov/node/1087. See also discussion in this report regarding dealing with uncertainties.

[20] Karl, Thomas R., Mellillo, Jerry M., and Peterson, Thomas C. (eds.) Global Change Impacts in the United States. U.S. Global Change Research Program. 2009. Such periodic assessments are required by the Global Change Research Act of 1990 (P.L. 101-606). A new national assessment of impacts on the United States is due in late 2013.

[21] There is a growing set of research on the relationship between past climate change and civilizations. A sample of recent research includes Buckley, B. M., K. J. Anchukaitis, D. Penny, R. Fletcher, E. R. Cook, M. Sano, L. C. Nam, A. Wichienkeeo, T. T. Minh, and T. M. Hong. "Climate as a Contributing Factor in the Demise of Angkor, Cambodia." Proceedings of the National Academy of Sciences 107, no. 15 (March 2010): 6748–6752; Cook, Edward R, Kevin J Anchukaitis, Brendan M Buckley, Rosanne D D'Arrigo, Gordon C Jacoby, and William E Wright. "Asian Monsoon Failure and Megadrought During the Last Millennium." Science (New York, N.Y.) 328, no. 5977 (April 23, 2010): 486– 489; DeMenocal, PB. "Cultural Responses to Climate Change During the Late Holocene." Science (Washington) 292, no. 5517 (April 27, 2001): 667–673; Haug, G. H, D. Gunther, L. C Peterson, D. M Sigman, K. A Hughen, and B. Aeschlimann. "Climate and the Collapse of Maya Civilization." Science 299, no. 5613 (2003): 1731; Scholz, Christopher A., Thomas C. Johnson, Andrew S. Cohen, John W. King, John A. Peck, Jonathan T. Overpeck, Michael R. Talbot, et al. "East African Megadroughts Between 135 and 75 Thousand Years Ago and Bearing on Early-modern Human Origins." Proceedings of the National Academy of Sciences of the United States of America 104, no. 42 (October 16, 2007): 16416–16421.

[22] For examples of these risks to power plants, see Department of Energy (DOE), U.S. Energy Sector Vulnerabilities to Climate Change and Extreme Weather. July 2013. http://energy.gov/sites/prod/files/2013/07/f2/20130710-EnergySector-Vulnerabilities-Report.pdf. See also CRS Report R43199, Energy-Water Nexus: The Energy Sector's Water Use, by Nicole T. Carter.

[23] NRC Committee on the Development of an Integrated Science Strategy for Ocean Acidification Monitoring, Research, and Impacts Assessment; National Research Council. "Executive Summary." In Ocean Acidification: A National Strategy to Meet the Challenges of a Changing Ocean. Prepublication. Washington, D.C.: The National Academies Press, 2010; Jacobson, Mark Z. "Studying ocean acidification with conservative, stable numerical

schemes for nonequilibrium air-ocean exchange and ocean equilibrium chemistry." Journal of Geophysical Research 110 (April 2, 2005): 17 PP.

[24] Ibid.

[25] An example of this is the adverse weather events in early 2011 that led to spikes in food prices and contributed to demonstrations in Tunisia and Egypt. These, in turn, led to regime change, although one cannot attribute these events to climate change, as opposed to weather variability, and the political implications might have been very different in regimes with better economic performance, less income disparity, fewer allegations of corruption, and greater social resilience. The point remains, nonetheless, that societies are sensitive to climatic variables in many ways.

[26] Regarding risks to national security, see, for example, Defense Science Board Task Force on Trends and Implications of Climate Change for National and International Security. October 2011. http://www.acq.osd.mil/dsb/ reports/ADA552760.pdf; and U.S. Department of Defense. Quadrennial Defense Review, February 2010. (pp. xv, 84- 88) http://www.defense.gov/qdr/qdr%20as%20of%2029jan10%201600.pdf.

[27] Economists and scientists sometimes refer to "ecosystem services," which are the services that natural systems provide and for which, very frequently, humans do not typically pay. Ecosystems services include water filtration, filtering of air pollution, recreational and spiritual opportunities, etc. Even without being valued in capital markets, ecosystem services may be critically important to economies. For example, in many coastal areas, mangroves or wetlands provide valuable buffering against frequent storm and flood events. If such ecosystem services did not exist, communities would have to pay for manufactured alternatives (e.g., sea walls) or risk incurring damages.

[28] For example, the very rapid spread of pine beetles in recent years was unexpected and caused large damages (although a temporarily inexpensive supply of timber) in a very short period. See CRS Report R40203, Mountain Pine Beetles and Forest Destruction: Effects, Responses, and Relationship to Climate Change.

[29] Malcolm, Jay R., Canran Liu, Ronald P. Neilson, Lara Hansen, and Lee Hannah. "Global Warming and Extinctions of Endemic Species from Biodiversity Hotspots." Conservation Biology 20, no. 2 (2006): 538-548.

[30] John W. Williams, Stephen T Jackson, and John E. Kutzbach, "Projected distributions of novel and disappearing climates by 2100 AD," Proceedings of the National Academy of Sciences of the United States of America 104, no. 14 (April 3, 2007).

INDEX

#

20th century, 89, 94, 104
21st century, 97

A

abuse, 77
access, 2, 3, 18, 32, 35, 42, 47, 51, 54, 58, 60, 61, 63, 64, 70
accessibility, 59, 61
accountability, 10
accounting, 10, 81
accreditation, 65
acidity, 90, 101, 104
AD, 106
adjustment, 76
advancement, 83
adverse effects, viii, 88
adverse event, 10
adverse weather, 106
aerosols, 90, 93, 94
aerospace, 9, 28
age, 44
agencies, viii, 2, 4, 5, 11, 12, 32, 34, 36, 48, 55, 56, 57, 58, 59, 60, 63, 65, 66, 68, 69, 70, 71, 72, 73, 76, 78, 80, 83, 84, 85, 101
agriculture, 56, 90, 93, 98
Air Force, 30, 31
Alaska, 11
alternative causes, 91
American Recovery and Reinvestment Act, 4, 13
American Recovery and Reinvestment Act of 2009, 4, 13
appropriations, 13, 60, 78, 92
Appropriations Act, 78
assessment, 15, 20, 32, 44, 56, 57, 68, 76, 77, 78, 80, 82, 83, 84, 103, 105
assessment tools, 83
assets, 6, 9, 12, 17, 30, 32, 44, 45, 46, 57, 73, 83
atmosphere, 5, 90, 92, 93, 95, 96, 101, 104
attribution, 91
audit, 7, 76
authority(s), 6, 60, 78, 92
awareness, 56, 83

B

banking, 81
barriers, 48, 51
base, 30
beetles, 106
benefits, viii, 3, 33, 44, 67, 68, 70, 77, 81, 88, 90, 105
bias, 29, 49
biodiversity, 102
bounds, 95, 100
browser, 18

businesses, 16, 56

C

Cambodia, 105
candidates, 74
capital markets, 106
carbon, 90, 93, 96, 97, 101, 104
carbon dioxide, 90, 93, 97, 101, 104
case study(s), 7, 62, 67, 74
category a, 15, 81
category b, 74
challenges, 3, 4, 5, 6, 10, 13, 31, 32, 33, 34, 36, 37, 38, 46, 50, 59, 80, 102
China, 93
cholera, 101
circulation, 49
cities, 105
civil servants, 28
civil service, 30
civil society, 31
Clean Air Act, 82
climate change issues, 31, 53
climate conditions, vii, 2, 44, 47, 50, 95
climate shifts, viii, 87
closure, 19
clouds, viii, 87, 95
CO_2, 90, 93, 95, 96, 101, 104
Coast Guard, 30
coherence, 59
collaboration, 36, 46, 60, 75
commodity, 99
communication, 36, 64
community(s), 3, 5, 21, 25, 35, 46, 47, 53, 56, 62, 64, 65, 67, 71, 101, 102, 106
complexity, 11
compliance, 75
complications, 81
composition, 92
computer, 18
conditioning, 46
conference, 68
conflict, 6
Congress, 68, 76, 87, 92, 101, 103, 104
Congressional Budget Office, 8, 74, 77
connectivity, 42
consensus, 66, 71, 77, 90
conservation, 102
consolidation, 58, 61
construction, 13, 23, 42, 43, 47, 51, 54, 67
controversial, viii, 87
controversies, 88
convergence, 16
cooling, 24, 89, 100, 101, 105
coordination, 58, 59, 61
Copyright, 18
coral reefs, 101
corrosion, 21
corruption, 106
cost, 5, 8, 9, 39
Court of Appeals, 82
crises, 3
critical analysis, 56
critical infrastructure, 4, 63
customers, 53, 79
cycles, viii, 24, 33, 87, 104
cyclones, 5

D

damages, 4, 5, 11, 92, 106
data collection, 58
database, 54, 64
decision makers, vii, 1, 2, 3, 6, 7, 11, 12, 32, 33, 34, 35, 36, 37, 38, 46, 47, 50, 51, 52, 53, 54, 56, 58, 59, 60, 61, 62, 63, 64, 65, 66, 70, 71, 72, 73, 74, 75, 77, 80, 84
decision-making process, 34
deficiencies, 10
deforestation, 90, 93
degradation, 93
demonstrations, 106
Department of Defense, 3, 6, 76, 106
Department of Energy, 79, 105
Department of Homeland Security, 6, 19, 76, 79
Department of Transportation, 7, 40, 42, 48, 51, 52, 54, 62, 63, 71, 72, 75, 78, 79, 80, 81, 83, 84, 85
Departments of Agriculture, 76

designers, 41, 54
diarrhea, 101
disaster, 6, 81
discharges, 80
diseases, 99
distribution, 25, 102
District of Columbia, 82
divergence, 12
diversity, 7, 38, 56, 65
DOT, 43
draft, 57, 63, 68, 69, 71, 72, 85
drainage, 15, 42
drinking water, 24, 62, 78
drought, 5, 20, 47, 100
drying, 98
durability, 39

E

ecological systems, 97, 102
economic growth, 8
economic losses, 102
economic performance, 106
economics, ix, 88, 91
ecosystem(s), viii, 88, 90, 91, 96, 101, 102, 105, 106
education, 65
effluent, 20
Egypt, 105, 106
El Niño, 96
emergency, 24, 39
emergency relief, 39
Emergency Relief Program, 78, 81
emergency response, 24
emission, 96
employees, 30
employers, 65
energy, 13, 14, 28, 31, 56, 77, 85, 90, 93, 98, 99, 105
energy conservation, 13
engineering, 36, 37, 38, 52, 65, 67, 71, 76, 77, 81
environment, 8, 10, 11, 31, 33, 42, 56, 68, 76, 83, 84, 85
environmental conditions, 76
environmental effects, 54, 68
environmental impact, 15, 68, 84
environmental impact statement, 68, 84
environmental management, 54
Environmental Protection Agency (EPA), 3, 7, 8, 9, 13, 16, 20, 47, 51, 59, 62, 63, 71, 72, 74, 75, 76, 77, 78, 79, 80, 82, 83, 84
environmental stress(s), 5
equilibrium, 106
equipment, 44
erosion, 79, 101
evacuation, 19
evapotranspiration, 98
evidence, viii, 7, 76, 82, 88, 91, 103, 104
exclusion, 84
executive branch, 55, 69
executive order(s), 47, 48, 55, 69, 82
expertise, 3, 36, 52, 64, 65, 73
exploitation, 100
extreme cold, 98
extreme precipitation, 24, 37
extreme weather events, 5, 6, 14, 15, 19, 47, 52, 57, 61, 67, 69, 99

F

federal agency, 74
federal assistance, 8
federal facilities, 5, 8, 10, 12, 31
federal funds, 13, 78
federal government, vii, 1, 3, 5, 6, 8, 9, 12, 55, 56, 57, 58, 59, 60, 63, 64, 65, 66, 69, 75, 78
Federal Highway Administration (FHWA), 7, 13, 57, 63, 66, 69, 74, 75, 78, 83, 85
fertilization, 90
filtration, 106
financial, 9, 46, 59, 92, 101
financial incentives, 92
financial markets, 101
financial resources, 46
fish, 42
fisheries, 101
flexibility, 13
flight, 25, 28

flooding, 5, 14, 16, 17, 20, 21, 24, 28, 30, 31, 41, 43, 54, 79, 80, 98
floods, 5, 16
fluctuations, 97
food, 101, 106
force, 55, 56, 64, 82, 93
formula, 12
foundations, 16, 42
fraud, 77
funding, 6, 8, 10, 12, 13, 30, 33, 41, 59, 60, 69, 73, 77
funds, 4, 13, 39, 69, 78

G

GAO, vii, 1, 2, 3, 22, 24, 40, 43, 73, 76, 77, 78, 79, 83, 85
General Services Administration, 9, 77
geographic scales, viii, 87, 90
GHG, viii, 87, 88, 90, 91, 93, 94, 95, 96, 97, 98, 104
GHG concentrations, viii, 87, 88, 90, 91, 93, 94, 97
GHG emissions, viii, 88, 90, 91, 93, 97
Global Change Research Act, 56, 76, 82, 105
global warming, viii, 48, 88, 90
governance, 102
governments, 8, 66
grant programs, 13
grants, 12, 77
gravity, 28
greenhouse, viii, 5, 29, 49, 82, 85, 87, 91, 93
greenhouse gas(s) (GHG), viii, 5, 29, 49, 82, 85, 87, 91, 93
greenhouse gas emissions, 49, 82
gross domestic product, 19
groundwater, 24
growth, 95
guidance, 7, 11, 12, 52, 60, 63, 66, 67, 68, 69, 71, 72, 73, 85
guidelines, 78, 81
guiding principles, 55
Gulf Coast, 17, 28, 63, 75, 79, 84
Gulf of Mexico, 17, 18, 25, 41

H

habitat(s), 24, 42, 102
Hawaii, 57
hazards, 39, 46
health, 24, 76, 82
Health and Human Services, 76
health problems, 24
heat-trapping gases, 14
height, 22, 81
high winds, 30
highways, 9, 13
historical data, 12, 37, 38, 51
history, 89
Holocene, 104, 105
homes, 16, 19
hotspots, 102
House, 4, 104
House of Representatives, 4
human, vii, viii, 1, 5, 10, 25, 28, 56, 88, 90, 91, 93, 97, 98, 100, 102, 104
human health, 56
Hurricane Katrina, 17, 18, 39, 47, 52, 81
hurricanes, 14, 17, 30, 39, 98
hybrid, 100
hypothesis, 91

I

improvements, viii, 80, 88
income, 93, 106
increased access, 55
individuals, 35, 48, 74, 101
industry, 19, 46, 74, 78, 90, 93
information technology, 12
infrastructure projects, viii, 2, 8, 12, 13, 32, 39, 66
inspectors, 74
institutions, 3, 6, 7, 36, 46, 58, 64, 70, 74, 75
insurance policy, 8
integration, 61

integrity, 15, 16
interface, 65, 77
interference, 102
intermediaries, 35
International Space Station, 9, 25
investment(s), 8, 10, 11, 12, 33, 39, 51, 66, 67, 68, 69, 70, 71, 81, 100
irrigation, 100
islands, 17
issues, vii, 1, 12, 38, 56, 63, 83

K

Kyoto Protocol, 95, 96

L

Lake Pontchartrain, 17, 39, 40, 52
land cover, viii, 87, 90
laws, 7, 73
lead, 6, 11, 17, 21, 42, 71, 95, 97, 102
leadership, 25, 47
leaks, 93
learning, 3, 62
Lebanon, 105
legislation, 60, 67
levees, 100
life cycle, 12
light, 31, 38, 92
loans, 9
local authorities, 12
local community, 52
local government, 8, 10, 12, 33, 46, 70, 78
longevity, 65
Louisiana, 2, 7, 17, 18, 19, 39, 40, 41, 42, 52, 74, 75, 79, 81
lying, 16, 17, 19, 20, 31

M

magnitude, 8, 11, 37, 47
majority, viii, 31, 37, 88, 91, 93
management, 5, 6, 7, 8, 9, 10, 11, 12, 15, 20, 21, 24, 32, 44, 46, 54, 61, 62, 66, 69, 72, 74, 78, 79
mangroves, 106
marine species, 101
marsh, 46
mass, 104
materials, 14, 24, 49, 67, 75
materials science, 67
mathematics, 91
measurements, 91
media, 88
mediation, 64
medical, 25
melt, 16
melting, 97, 98
mental health, 102
methodology, 7, 81
Mexico, 18
migration, 99, 102
military, 8, 31, 32
misconceptions, 104
mission(s), 6, 25, 28, 31, 32, 46, 48, 56, 59, 60, 73, 76
Mississippi River, 79
models, 28, 29, 30, 35, 36, 49, 52, 56, 58, 61, 65, 81, 98, 101
modifications, 43, 44
moisture, 15
mortality, 100

N

National Aeronautics and Space Administration (NASA), vii, 1, 2, 3, 5, 6, 7, 9, 15, 24, 25, 26, 27, 28, 29, 30, 31, 32, 45, 46, 47, 48, 49, 50, 52, 53, 54, 62, 66, 69, 72, 74, 76, 77, 79, 80, 82
National Academy of Sciences, 76, 89, 105, 106
national borders, 102
national interests, 6
National Research Council, vii, 2, 4, 5, 73, 101, 103, 104, 105
national security, 6, 106

national strategy, 55, 80, 81, 82, 104, 105
natural disaster(s), 47, 67, 78
natural resources, 30, 101
nerve, 28
next generation, 30
nitrous oxide, 90, 93
NOAA, 4, 5, 14, 18, 19, 36, 38, 49, 50, 51, 55, 60, 64, 65, 76, 80, 81, 84, 101, 104
nonequilibrium, 106
NRC, vii, 2, 3, 4, 5, 6, 8, 10, 14, 15, 16, 24, 32, 33, 34, 35, 37, 38, 47, 48, 53, 56, 57, 58, 60, 63, 64, 65, 67, 70, 76, 77, 78, 79, 80, 81, 82, 83, 84, 101, 105

O

oceans, viii, 87, 90, 101, 104
Office of Management and Budget, 4, 78
Office of the Inspector General, 77
officials, 2, 6, 7, 13, 16, 18, 19, 20, 23, 25, 26, 32, 33, 34, 39, 40, 41, 42, 44, 46, 47, 48, 51, 52, 53, 54, 57, 58, 59, 60, 61, 62, 63, 66, 67, 73, 74, 75, 76, 78, 79, 80, 84
oil, 2, 18, 19
Oklahoma, 75
OMB, 4, 12, 60, 61, 66, 71
openness, 47
operations, 18, 25, 48, 55, 56, 73
opportunities, vii, 1, 6, 7, 11, 54, 57, 58, 69, 73, 75, 78, 101, 102, 106
orbit, 90, 93, 104
organize, 59
oscillation, 96
outreach, 36, 53, 63, 83
outreach programs, 53
oversight, 83
ozone, 93

P

Pacific, 14, 75, 84, 96
participants, 52, 63
payroll, 28
peer review, 34, 74
permafrost, 104
permit, 33, 80, 84
pH, 101
planning decisions, 11
plants, 9, 20, 21, 22, 33, 80
policy, 47, 48, 55, 56, 59, 67, 69, 77, 82, 100
policy making, 77
policy options, 77
policymakers, 5, 11
pollutants, viii, 20, 87
pollution, 5, 99, 106
ponds, 42
population, 5, 38
population growth, 5, 38
portfolio, 9
positive feedback, 95, 104
power plants, 105
precipitation, vii, viii, 2, 5, 14, 15, 16, 20, 23, 24, 28, 30, 31, 37, 38, 41, 49, 50, 79, 80, 81, 88, 95, 97, 98, 99, 100
preparation, 53, 80
preparedness, 53
preservation, 81
President, 2, 55, 68, 71, 76, 85, 103
President Obama, 103
principles, 10, 11
private sector, 8, 59
probability, 23, 38, 79, 82
probability distribution, 38
producers, 34
professionals, 25
profit, 101
project, viii, 12, 13, 16, 23, 28, 30, 38, 42, 43, 46, 48, 49, 51, 54, 57, 64, 66, 68, 69, 70, 71, 74, 81, 84, 88, 90, 97, 102
protection, 99
public awareness, 54, 55
public health, 8, 20, 33, 82
public policy, 64, 76

Q

Quadrennial Homeland Security Review, 6, 76

quantification, 32

R

radiation, viii, 88, 93
rainfall, 3, 17, 23, 44, 47, 79
rating scale, 10
real property, 9, 25
real time, 81
reality, 11
recognition, 8, 65, 83
recommendations, 55, 57, 71
reconstruction, 16
recovery, 4, 5, 10, 81
recreation, 24
recreational, 44, 106
regulations, 7, 54, 68, 71, 73
rehabilitation, 9, 43, 54
reliability, 81, 99
repair, 9, 10, 16, 70
requirements, 13, 33, 66, 68, 71, 80
researchers, 34, 36, 51, 102
reserves, 102
resilience, 10, 40, 42, 56, 71, 79, 101, 106
resolution, 29, 49, 77
resource management, 64
resources, 2, 10, 11, 24, 32, 55, 59, 62, 63, 83, 100, 101
response, vii, 1, 5, 6, 10, 11, 47, 53, 60, 76, 92, 104
restoration, 46
restrictions, 16
rights, 18
risk(s), vii, ix, 1, 3, 4, 5, 8, 10, 11, 12, 16, 17, 20, 21, 23, 25, 28, 32, 38, 39, 44, 45, 46, 47, 48, 52, 53, 54, 56, 57, 62, 67, 69, 70, 71, 76, 78, 79, 83, 88, 91, 92, 97, 98, 100, 101, 102, 105, 106
risk assessment, 32, 57, 62, 78, 83
risk management, 5, 8, 11, 12, 38, 70, 78
river flows, 16
routes, 16
rules, 33, 80, 82
runoff, 3, 16, 21, 24, 44, 46, 79, 98, 100

S

safety, 19, 33, 42, 51, 81
salinity, 24
saltwater, 21, 40, 44
saturation, 42
science, viii, 5, 35, 36, 52, 53, 55, 56, 57, 59, 61, 62, 64, 65, 67, 70, 76, 77, 79, 81, 87, 88, 90, 91, 92, 97, 103, 104
scientific knowledge, 82
scientific method, viii, 88, 91
scientific papers, 34
scientific theory, 91
scientific understanding, 94, 98
scope, viii, 7, 73, 83, 88, 90
sea level, vii, viii, 2, 5, 8, 14, 15, 16, 17, 18, 19, 20, 21, 23, 25, 28, 30, 31, 32, 41, 43, 45, 49, 51, 52, 68, 78, 85, 87, 88, 97, 98, 99, 100
sea level rise, viii, 2, 5, 8, 14, 15, 16, 17, 19, 20, 21, 23, 28, 30, 31, 32, 41, 43, 45, 49, 51, 68, 78, 85, 87, 88, 97, 98
seasonality, 81
security, 31, 102
sediment, 16
Senate, 4, 67
services, 18, 55, 59, 61, 79, 106
sewage, 21, 79, 81
shape, 31
shoreline, 24, 25
short supply, 35, 101
signs, 41
simulation(s), 28, 31, 49, 61
society, 11, 76, 99, 101
software, 62
species, 24, 99, 101, 102
specific knowledge, 74
spending, 8
Spring, 55, 80
stability, 102
staffing, 33
stakeholders, vii, 2, 3, 6, 7, 25, 30, 34, 36, 37, 46, 47, 51, 54, 57, 58, 60, 61, 64, 66, 70, 71, 73, 75, 102

state(s), 5, 8, 10, 11, 12, 13, 16, 18, 19, 31, 33, 34, 35, 39, 40, 41, 42, 47, 48, 50, 51, 57, 60, 63, 64, 66, 67, 68, 69, 70, 74, 75, 78, 79, 83, 85, 97
statistics, 65
statutory authority, 78
stock, 73
storage, 23, 43, 101
storms, viii, 14, 15, 17, 20, 21, 24, 30, 39, 79, 88, 98, 99
stormwater, 8, 16, 24, 33, 42, 44, 62, 78, 80
strategic planning, 12, 56, 64, 70
stress, 24, 31, 102
stretching, 50
structure, 39, 40, 41, 101
sulfur, 93
Sun, viii, 87, 93
sustainability, 82, 85

T

tanks, 74
Task Force, 55, 65, 82, 106
teams, 48, 64
technical assistance, 36, 58, 61, 62, 63
technical comments, 2, 16, 20, 37, 38, 51, 56, 60, 72, 78, 79, 80, 81, 82, 83, 84, 85
technological advances, 100
technology, 61, 76, 79, 82, 92
temperature, 5, 14, 15, 16, 28, 30, 37, 49, 50, 52, 89, 90, 91, 94, 97, 98, 99, 101, 104, 105
tenure, 36
term plans, 33
testing, 25, 30
the National Oceanic and Atmospheric Administration's (NOAA), 5
threats, 28, 30, 45
tides, 2, 19
time frame, 6, 33, 68, 75
time periods, 104
time series, 89, 99
tracks, 14, 16, 52
trade, 8, 74
training, 25, 32, 65
transformation, 77
translation, 62, 64, 65
transport, 79
transportation, 4, 8, 13, 16, 17, 33, 39, 40, 41, 42, 47, 48, 51, 56, 57, 60, 63, 66, 67, 68, 70, 74, 78, 83, 84, 85, 99, 100
transportation infrastructure, 47, 57, 60, 67, 68
treatment, 4, 8, 9, 20, 21, 33, 42, 79, 80
tribal officials, 5
tropical storms, 17

U

U.S. Department of Agriculture, 64
U.S. Department of Commerce, 89
UK, 104, 105
uniform, 89
United, v, 1, 4, 5, 8, 9, 14, 15, 25, 26, 31, 50, 55, 71, 73, 75, 76, 78, 79, 81, 82, 83, 89, 93, 96, 98, 103, 105, 106
United States, v, 1, 4, 5, 8, 9, 14, 15, 25, 26, 31, 50, 55, 71, 73, 75, 76, 78, 79, 81, 82, 83, 89, 93, 96, 98, 103, 105, 106
universe, 74
universities, 34
updating, 67
urban, 44, 93
urbanization, 5, 21

V

vapor, 104
variables, 37, 106
variations, 97, 98
varieties, 100
vector, 24
vegetation, viii, 44, 87, 90, 93, 95
vehicles, 16, 82
velocity, 79
ventilation, 46
vulnerability, 10, 15, 17, 21, 23, 25, 31, 44, 46, 51, 52, 55, 57, 60, 63, 73, 78, 83, 101

W

Washington, 7, 16, 21, 41, 43, 48, 51, 53, 54, 57, 64, 67, 74, 75, 76, 77, 78, 79, 80, 81, 82, 83, 84, 104, 105
waste, 77
wastewater, vii, 1, 2, 4, 5, 6, 7, 8, 9, 10, 12, 15, 20, 21, 23, 32, 33, 47, 52, 54, 62, 66, 69, 70, 71, 72, 74, 78, 79, 80
wastewater systems, vii, 1, 2, 6, 71, 73
water, 8, 9, 12, 13, 20, 21, 41, 42, 43, 44, 46, 54, 56, 57, 62, 77, 79, 80, 83, 84, 85, 90, 93, 95, 98, 99, 100, 101, 104, 106
water quality, 9, 20, 44, 62, 84
water quality standards, 20
water resources, 56, 99
water vapor, 90, 93
watershed, 80
waterways, 16, 42
web, 67, 79
websites, 73
welfare, 56, 82
wetland mitigation, 43
wetland restoration, 81
wetlands, 3, 106
White House, 85
wildlife, 42, 43
wind gusts, 30
Wisconsin, 7, 23, 24, 44, 51, 59, 74, 75, 79
workers, 19
workforce, 26, 65
worldwide, 31, 77, 78